Black Swamp Farm

Black
Swamp
Farm

by Howard E. Good

Illustrations by Brenda Olson Sutherland

Ohio State University Press

Library of Congress Cataloging-in-Publication Data

Good, Howard E.
    Black Swamp farm / by Howard E. Good; illustrations by Brenda
Olson Sutherland.
        p.   cm.
    ISBN 0-8142-0734-0 (paper: alk. paper)
    1. Farm life—Ohio.   2. Good, Howard E.   3. Farmers—Ohio—
Biography. I. Title.
S521.5.O3G66   1997   96-48331
977.1'12041—dc21   CIP

Cover design by Mike Jaynes.

The paper used in this publication meets the minimum requirements of the
American National Standard for Information Sciences—Permanence
of Paper for Printed Library Materials. ANSI Z39.48-1992.

9   8   7   6   5   4   3   2

*In memory of
my father and mother
and my wife*

Preface

THIS IS A FACTUAL ACCOUNT, intended to present a faithful portrayal of farm life in the old Black Swamp region of the Maumee Valley in northwestern Ohio, as experienced and witnessed by one who was born in a log house on a farm there late in the last century and spent there more than a score of his earlier years.

During most of the years in which I have been away from that farm, my home has been on the shore of the Maumee River, not in, but near, the Black Swamp. Hundreds of times I have driven into or through that region, thinking often of the original Indian inhabitants and the eminently valid reasons they had—many of those reasons before me in the lovely vistas of the valley—for resisting stubbornly and fighting fiercely against the whites who came to dispossess them. Twice or oftener each year, I have returned to the environs of the old home, taking note of developments and changes that come unfailingly with the march of time.

White men who first came to northwestern Ohio found all of it very wet during much of each year. The boggiest portion, lying to the south and east of the Maumee River, came eventually to be named the "Black Swamp." Because the wetness and the extent of the swamp varied from season to season and from year to year, scarcely any two observers agreed in defining its boundaries. For practical purposes it may be described as an area thirty to forty miles in width, approximately parallel to the river, extending from Lake Erie southwestward to a point a few miles beyond the Ohio-Indiana line.

Ice age conditions made the Black Swamp what it was and is, giving to it an interesting combination of characteristics in all probability not duplicated exactly anywhere in the world. Ages ago the region was left overlain by a glacial drift sheet of table-like flatness on which grew a great expanse of forest. Because of deficient natural drainage, water from rains and snows stagnated over the land during untold centuries, forming a huge dark swamp.

Reclamation by drainage and clearing of forest growth, continuing altogether about a full century, involved an incredible amount of toil and the expenditure of an enormous grand total of money. This must be rated an achievement of heroic proportions.

While the narrative covers only a relatively short period of personal observation—daily from about 1892 until 1906, somewhat intermittently thereafter—it harks back to the days of brave, hardy men and women who refused to yield to discouragement or despair in the face of severe hardships and grueling toil. It also recalls conditions, problems, mores, and concepts that are beginning to fade into the shadowy realm of the legendary and the forgotten.

The farm was sold in 1946, my mother having died and my aged father being then no longer able to remain there.

viii

It is well kept by its present owner, who operates it in conjunction with neighboring lands, making for him a total of some five hundred acres. It remains among the best and most productive in the community.

Strikingly illustrative of the changes that have come about since our early days there (when, if one fared abroad in early spring, he had to walk, ride a horse, or depend on a two-wheeled horse-drawn cart to get through the jack-wax of unpaved roads), an airplane now facilitates travel from and to that farm. A landing strip and hangar are maintained in a field where we often labored with hoes in the heat of summer and, at times, ate melons fresh from the vines, or "whipped" potato bugs with small branches from a tree.

HOWARD E. GOOD

Contents

The Swamp and How It Was Made   1   One

Stouthearted Men with Axes   13   Two

Drainage   21   Three

The Soil   29   Four

Above and Beyond the Call of Duty   35   Five

The Well-fed Farm Family   47   Six

Our Home on the Farm   61   Seven

A New Barn and a New House   71   Eight

Rescued from Seas of Mud   89   Nine

Animal Companions and Aids   97   Ten

The Beginning of the Farmers' Year   117   Eleven

Farmers' Payday   125   Twelve

The Fall: Full Job Quotas   141   Thirteen

Work and Play for Farm Boys   151   Fourteen

Prevailing Fashions   165   Fifteen

Callers at Farmers' Doors   177   Sixteen

The People of Black Swamp   181   Seventeen

Language of Everyday Communication   195   Eighteen

North School, District No. 10   205   Nineteen

The Fair   233   Twenty

Diversions and Entertainment   241   Twenty-one

Medicines: Amateur and Professional Doctors   255   Twenty-two

Superstitions and Tales   269   Twenty-three

Far-reaching Changes   281   Twenty-four

Recent Looks at the Valley   291   Twenty-five

Black Swamp Farm

ONE

The Swamp and How It Was Made

THE HISTORICAL RECORDS,
though somewhat fragmentary and indirect, indicate that
the French explorer Robert Cavelier de La Salle and a
group of his compatriots were the first of the white race to
behold the Maumee Valley, in the northwestern quarter
of what is now the state of Ohio. That was about the year
1669. In ensuing years other Frenchmen—explorers, mis-
sionaries, and traders—visited the region. At the urging of
the colonial governor, most of them kept ever in sight the
objective of strengthening and confirming the claim of the
French crown to a vast interior area of the continent, of
which this valley was a part. The British, who considered
this territory rightfully theirs, viewed these incursions with
an extremely jealous eye; it was not long until they had
men on the scene, maneuvering to checkmate the French.

No doubt all these Europeans, viewing the valley in its
pristine state, marked certain features that set it apart as
different from anything they had seen elsewhere. For one
thing, they found the native population relatively large.
The Indians had a particular liking for the region and
tended to concentrate in it because its rich natural re-

sources supplied them abundantly with the necessities of life with only a minimum of effort on their part.

Moreover, it had for them no mean strategic value, being traversed by one of the most convenient and useful routes of travel in their far-flung domain. Via the Maumee, the Wabash, the Ohio, and the Mississippi rivers, the red men could move easily by canoe in either direction between the Great Lakes and the Gulf of Mexico with only simple, minor portages to impede them. These rivers, the Great Lakes, the St. Lawrence, and streams tributary to them gave them ready access to nearly all of the continent east of the Rockies.

The whites could hardly have failed to note that the land was extraordinarily flat, much of it lying at almost dead level. Nowhere was there an elevation that could qualify even as a hill. Over thousands of acres countless trees that towered to unusual heights grew in dense, unbroken stands. The big trees stood like straight, giant poles, branches and leaves joining and intertwining at the tops to form far above the ground a vast canopy that almost completely shut out the sun's rays.

In the woods they found innumerable deer. There were also many bears, wolves, wildcats, foxes, and the like, as well as a wide variety of game, songbirds, and birds of prey. Snakes, including deadly rattlers, moccasins, and copperheads, were likewise in evidence. Fish abounded in the streams. Armadas of bloodthirsty mosquitoes lurked everywhere, and on every hand frogs and turtles were present in great numbers.

The palefaces encountered great reaches of water lying, dark and motionless, over much of the flat terrain. This stagnant accumulation, in combination with the waxy gumbo beneath it, made travel other than by boat or canoe in the streams extremely difficult and unpleasant, not to say hazardous at times. Eventually, as the region became familiar to more and more white men, the boggiest and wettest portion, a belt lying along the Maumee River, principally to the south and east, thirty to forty miles wide and about 120 miles long, came to be generally known as the "Black Swamp."

Bearing in mind the great expanses of murky water, the sticky mire, and the fearful darkness of the deep, dense woods—to say nothing of the Indians, the snakes, and the predaceous animals—it must be admitted that this was a very appropriate name. For several recent generations, however, "Black Swamp" has been very decidedly a misnomer. As a matter of fact, the general character of the region has been so greatly altered that a considerable number of the many people now living in it are only vaguely aware—if aware at all—that their homes are in what a long time ago was a great waterlogged swamp, so dark, so forbidding, that it could be fittingly described as "black."

How did this region become the Black Swamp? How was it eventually transformed so that today the casual observer can scarcely believe that only a relatively short time ago it was a heavily forested wilderness, much of it under water, replete with wildlife and a favorite retreat of the red man?

It is a long story. To start the account somewhere near the real beginning—that is, with that period in our planet's history that ultimately saw the valley assume the character it had when white men first viewed it—it will be necessary to "flash back," as they say in movie studios. We shall have to go back a long way, through thousands of centuries. Some of the geologists who have studied available data are of the opinion that the period we are considering began half a million years ago. There are those who argue that this estimate is too low, others, that it is too high. For our purposes here, however, quibbling over some thousands of years, more or less, would be pointless and, in large measure, bootless.

It is an established fact that at some time in the distant past masses of snow began to collect in unprecedented quantities in areas below the Arctic Circle, lower layers of snow becoming solid ice as weight increased. How or why all this happened not even learned men of science can say with certainty. Whenever it may have occurred, whatever condition or combination of conditions may have brought it about, the accumulation of that snow and ice was the opening phase of a most amazing natural phenomenon.

Slowly and irregularly, during a very long period, the ice piled up higher and higher. At length the accumulation attained such volume and such weight that enormous masses, like vast rivers, began to flow out from the centers of glaciation in both the Eastern and Western Hemispheres. The formation and initial flow of those ice masses, or glaciers, marked the onset of the earth's ice age, one of the most striking and momentous periods in its relatively recent history.

From studies of existing glaciers it is known that that ice flow was never fast, never regular; the rate of movement was directly proportional to the snow accumulation, which varied widely from year to year. The flow was on a gigantic scale, and its force was irresistible. The ice picked up and carried with it rock fragments that ranged in size from boulders almost as large as a house down to small pebbles. Measureless masses of sand and clay were accumulated and moved along in the ice streams. This burden of materials, all abrasive, vastly increased the power of the glaciers to cut and grind and crush. Mountains and hills were gashed and furrowed as if by some colossal engine of limitless power. Hard rocks of the earth's outer crust were deeply grooved and abraded. (Examples of such rock carving have been found in many places. Among the most striking are those appearing in the hard limestone of Kelley's Island in Lake Erie, lying to the north of Sandusky, Ohio. Excavations there have revealed acres of bedrock plowed in smooth, parallel furrows and flutes. Numerous geologists and multitudes of tourists have visited this island and viewed the famed glacial grooves in a huge exposed rock. The straight, parallel furrows in this monolith, deep and regular in contour, are practically the same as might have resulted if it had been passed through some mechanical shaper of fabulous size and power.)

In the glacial period ice sheets ranging up to a mile and more in depth spread over the northern part of our continent at one time or another. They covered a large part of present Canada, all of what is now New England and New York, and much of the area now occupied by the states of New Jersey, Pennsylvania, Ohio, Indiana, Illinois, Missouri, Iowa, Michigan, Wisconsin, Minnesota, and the Dakotas. In Montana the Rocky Mountains blocked the flow of ice from fountainheads to the northeast, but mountain glaciers streamed down to supplement and extend the continental sheet.

5

The glaciers flowed in numerous streams and lobes. Time after time they advanced, then receded, each advance, each recession, continuing many thousands of years. Scientists have established that four separate ice ages occurred. On our continent the four glaciations are known as the Nebraskan, the Kansan, the Illinoian, and the Wisconsin, the names indicating the regions in which each left its trace most indelibly.

At length, many millenia after the first accumulation of ice in far northern regions began, annual snowfalls diminished by degrees, and the southward movement of the last vast ice masses subsided very gradually. Solar heat slowly melted the ice so that the enveloping sheet retreated northward. The rate of this withdrawal was slow and irregular, apparently never exceeding a few hundred feet per year.

The topography of areas traversed by the glaciers was changed beyond recognition. Gone were many of the old hills. On our continent a new chain of low, gently sloping hills and ridges made up of debris released from the leading face of the ice mass, now extended in a line, curving irregularly, from the Atlantic coast to the Rocky Mountains. This was the terminal moraine, which marked the extreme southern spread of the ice sheet. Northward and roughly parallel to this moraine a number of other ridges traversed the drift plain in sagging, irregular lines, often broken and disconnected. These, composed of the same material, are generally less massive than the terminal moraine. They were formed when the melting process slowed from time to time, causing a temporary halt in the retreat of the ice sheet.

Mountains, roughly sculptured, presented strange new profiles. Old lakes that had not been completely obliterated

6

showed marked transformations. New lake beds and new rivers appeared. Many old streams had been filled up and buried under drift; others had been greatly altered, being displaced and forced into new channels or compelled to find new outlets. Oceans of fresh water from the melting ice escaped, directly or through newly carved channels, to preglacial streams. In the flood plains of old streams terraces were built up from silt that the torrents, pouring swiftly toward the sea, washed down from the moraines. In some instances the rushing waters, laden with abrasive materials, cut long, deep chasms and gorges through hard, solid rock.

Some of the preglacial valleys became mere shallow elongated depressions. Others disappeared completely, giving place to comparatively level plains. Drift, composed of boulders, gravel, sand, and clay, covered the primordial terrain to depths ranging from a few feet to several hundred feet. All of the present state of Ohio, with the exception of about one-third of its area, in the southeast, has a deposit averaging about one hundred feet in depth.

After the ice sheet, retreating northward, reached the southern divide of the Laurentian Basin, a vast tract now occupied in part by the Great Lakes, water from melting ice accumulated along the face of the ice wall, forming several lakes. The overflow from these lakes, of course, could not escape northward; it could get away only through old watercourses that carried much of it finally to the Mississippi River. There was also a mighty flow eastward across the present state of New York to the Hudson River. Outlet streams, including the Minnesota, the St. Croix, the Illinois, the Wabash, and the Ohio rivers—even the Father of Waters itself—to this day present features that were carved and constructed by that gigantic, prolonged flow, heavily charged with sand and rock fragments.

The ceaseless wash and surge of waves built up beaches about those lakes. As the glacial field slowly receded, the lakes also pulled back by degrees, abandoning old beaches and forming new ones. At last the original lakes all disappeared, but they left behind them the beaches that clearly tell the story, in each case showing outlets as well as changing areas and shore lines.

One of those ancient bodies of water, named Lake Maumee, extended in the long ago across northwestern Ohio and a corner of Michigan, from what are now Lakes Erie and Huron, to the site of the present city of Fort Wayne, Indiana. It had an outlet to the Wabash River through which its water flowed to the Ohio and thence to the Mississippi. There was also some discharge to the Ohio through the Scioto and Miami rivers.

In the course of its long life this lake built up a series of beaches that are still plainly to be seen. The outermost and highest of these ancient beaches extends into Ohio from Michigan at a point near the town of Fayette, Ohio, toward the northwestern corner of the state. From there it runs southwesterly to Fort Wayne, Indiana. From Fort Wayne the line is traced southeasterly through the towns of Van Wert and Delphos, in Ohio. At Delphos it shifts to a northeasterly course and extends to Findlay. After making there an odd double fold, it continues generally northeastward. Intermediate beaches run inside this one, nearer the existing shore of Lake Erie. These beaches are known locally as "ridges" and bear such names as "Sand Ridge," "Sugar Ridge," etc. Lying relatively high, with good natural drain-

8

age, most of them, from the days of the aborigines until now, have been used as trails or roadways.

The drift plain of northwestern Ohio was left so flat that creeks and rivers, all "young" by geological standards, with only slight fall, flowed sluggishly. For that reason they were unable to dispose of water as fast as it came down in rains and snows. This in large measure was responsible for the excessive wetness that originally prevailed during much of the year. The heavy forest growth also played an important role in retarding drainage.

The ancient beaches and the glacial moraines in most instances constituted impassable barriers to natural drainage. Generally, surface water was able to get away only by making detours around them. For instance, but for the beaches and moraines, water carried by the Blanchard River, which has its beginning near Kenton, could have continued its northerly flow directly to Lake Erie, over a route of about fifty miles, with a fall of about four feet per mile. Instead, it had to cut a channel from a point near Findlay to the westward, roughly parallel to the obstructing ridges, meandering about fifty miles to the Auglaize River. Thence the water is carried northward fifteen miles to the Maumee River and on to Lake Erie, some fifty miles distant, the fall averaging about eighteen inches per mile.

Another river, the St. Joseph, originating in Michigan, carries water over a roundabout route totaling over 160 miles to Lake Erie via the Maumee, with a fall averaging about twelve inches per mile. But for the barriers established in the ice age, it could have flowed over a direct route about sixty miles long to the lake, with a fall of five feet per mile.

My father was born on a farm that his father had chopped out of the virgin forest in the valley. In young manhood, he became the owner of a farm, some four miles distant, on which he spent more than sixty years of his life and on which my brother, Carey, my sister, Bessie, and I were born and grew to maturity. Both farms were in the Black Swamp, but in our time these and all other farms thereabouts were neither black or swampy. None of us, for years, heard the name Black Swamp applied to any part of the region—we didn't hear it mentioned at all, for that matter. Only the flatness of the country and the dark, somewhat waxy soil remained as clues to its original condition.

Our farm (now owned by others) lies four miles southeast of the town of Van Wert. It is in Van Wert County (named for Isaac Van Wert, one of the three American militiamen who captured Major André, the British spy, in 1789). A few miles to the north runs the outermost of the old Lake Maumee beaches. This east-west ridge has long been known locally as "the Ridge," and our township, which it crosses, was named Ridge Township.

The Ridge follows a fairly regular course, with an elevation several feet higher than that of adjacent land. It became for white men, as it had long been for the Indians, a major artery of travel. Today, the Lincoln Highway, Ohio –U.S. Route 30, runs over it throughout the county and beyond in both directions. The first homes in the area were established on or near this old beach because it afforded a convenient means for ingress and egress, and because the relatively dry soil was better adapted to tillage than the perennially wet lands on either side. The width of the Ridge in the county averages about a quarter mile. It is

10

generally higher along its southern side, sloping off gently toward the north. It is composed of sand and gravel with an admixture of shells, just as one would expect in any beach. Pioneers who built their cabins on the elevation found good water when they sank wells to a depth of fifteen or sixteen feet.

TWO

Stouthearted Men with Axes

FOR A LONG TIME HOSTILE
Indians discouraged settlement in the valley. For many
prospective settlers, the heavy forest growth, water on and
in the ground, mosquitoes, and ague were probably no less
powerful deterrents than the Indians. Immigration began
in a small way about 1820, the Indians having relinquished
their claim to the greater part of the region under terms
of a treaty signed a few years before. For several decades
after it began the influx of settlers was slow.

The earliest whites who established themselves in this
wet wilderness depended for a living mainly on hunting
and trapping. After the hunters and trappers came men
with their families in covered wagons, driving their livestock before them. They aimed to buy land and condition
it for farming. Like those who preceded them, they had to
depend largely upon food materials obtained in the woods.
Some of these people came from settlements outside Ohio;
others moved in from older Ohio counties, to the east and
southeast. There was also a trickle of immigration from
Germany and other European countries. No doubt the
principal attraction for all was the low cost of the land. A

good many Indians were still in the area, but they were not disposed to disinter the hatchet that their chiefs had buried; in numerous cases very friendly relations prevailed between them and white settlers.

Travel in Ohio at that time involved many difficulties, hardships, and dangers. Wolves and other predatory animals still lurked in the woods. Even worse were lawless men who preyed upon travelers, stealing horses and, on occasion, murdering men for their possessions. Families coming in from the east were harassed in some instances by laborers on the National Road, then being extended westward in sections through Ohio. A favorite trick of these rough fellows was to fell a tree across the thoroughfare ahead of an approaching wagon, then demand a fee for the removal of the obstruction.

If unbridged streams could not be forded, it was necessary to build rafts and pole families, wagons, and livestock across. During much of the year, especially in the Black Swamp region, such roads as had been opened were appallingly bad, with deep ruts and formidable mudholes whose negotiation would tax the patience, courage, and skill of the most stouthearted and seasoned traveler. It was not uncommon for two or more families to pool resources and travel in a wagon caravan. This was a decided advantage in meeting many of the difficulties, but of course it gave no protection against the ubiquitous mosquitoes or prostrating attacks of malaria.

The first task of a newcomer, after building a rude log cabin and doing some rough work to aid drainage, was to clear some of the land of trees. This verily, on the whole, was an undertaking that bordered on the insuperable, one at which only stouthearted, robust men could have succeeded. Few today can appreciate the awesome immensity of that virgin forest or imagine the fears and forebodings its dark depths tended to engender. Historians tell us that even after some of the Indian trails had given place to crude wagon roads and fairly extensive clearings had been made, it was not uncommon for women and children—even grown men—to get lost in the big stands of timber.

We were told that Robert Gamble, a pioneer who settled on a tract adjacent to the one that later became our farm, went into the dark woods one evening, guided by the sound of a bell worn by one of his cows. He rounded up the little herd, intending to return the animals to his clearing. They stubbornly resisted all of his efforts, breaking away time after time and running back to the starting point. A neighbor, hearing the commotion, found that Gamble was actually trying to drive the cows directly away from their home quarters. The animals knew that, but the man, completely "turned around," found it hard to believe.

One spring day a child of six, daughter of a pioneer family, wandered along the path leading from her father's cabin on the Ridge into the forest. When the sun dropped below the treetops at the edge of the clearing, her parents, noting that she had not returned as expected, became greatly alarmed, for the woods still harbored bears, wolves, and wildcats. The father fired several shots from his musket as a distress signal. Soon, in response, his nearest neighbor appeared. All night and into the following day, their hearts filled with apprehension and dread, the two men searched among the trees over a wide radius but they found no trace of the child. Toward noon of the second day an Indian chief, accompanied by two braves, strode to the neighbor's cabin, the little girl, unharmed but with tear-stained and bramble-scratched face, asleep upon his shoulder. She told how she had wandered along, lured by ever brighter and more attractive flowers, until at last the path seemed to have disappeared. She ran wildly, now in this direction, now in that, becoming more and more confused. Finally, as the dark curtain of night closed over the forest, exhausted and paralyzed by fear, she raked leaves together beside a log, lay down, and cried herself to sleep. There, hours later, the Indians found her. This was an oft-told tale in our family, for its central figure, in maturity, became my father's mother.

The settlers felled the big trees that grew about their cabins and cut them into logs, reserving the best for splitting into fence rails. They piled brush and unneeded logs in enormous heaps, sturdy men and sturdy oxen toiling through many long days. Finally, they applied the torch and the great heaps, which included some of the finest timber that ever grew, went up in smoke.

Three or four weeks of the most strenuous labor were required for a good woodsman, felling a tree at a time, to clear an acre of hardwood. In some instances clearing was expedited by an ingenious method known as "slashing." Starting at the side of a timbered tract opposite that from which the wind most commonly blew, the slasher would chop notches in all of the trees in a swath about thirty feet wide across the tract, cutting trunks one-third to one-half through. Notches were so placed as finally to direct the fall of each tree to the leeward and toward the center of the "windrow." Near the windward end of the slashed strip, the notches were all made considerably deeper. Then, with a brisk wind blowing in a direction parallel to the swath, a big tree was chopped off at the windward side as a "starter," to fall against one or more of its deeply-notched neighbors. The latter were knocked over and in tumbling threw down others to the leeward. And so it went, until, within a few minutes, all in the swath had fallen prone in a terriffic, prolonged, awe-inspiring crash. A competent slasher could cut off about an acre per day; two experts working together commonly counted on slashing twenty acres of heavy timber in about nine days.

After lying where they fell for a year or more, the dry trees were quickly disposed of by burning on a wholesale scale. Neighbors came from settlements near and far to help with the logging and burning. These "logging bees" were frontier social affairs in which an enormous amount of hard work, hearty repasts, and unlimited lusty fun were combined, to the delight of all participants.

The woodsmen collected quantities of the ashes that remained after big clearing fires. They leached the ashes, then boiled the lye to evaporate the water. The resulting potassium salt had an important market value at stores and trading posts. Some settlements had "asheries" that bought ashes and processed them, selling the potash to eastern establishments manufacturing soap and glass.

To us today, with lumber so scarce and expensive, this large-scale destruction of fine timber seems a rank, outrageous waste. It did not appear so at the time. To these men the forest was an encumbrance that they had to get rid of so that they could make the land yield a living. Timber for a long time had no market value—there was no market. In time, after much good timber had been destroyed, canals and railroads, made markets available, and a few sawmills and plants for the manufacture of barrel staves and hoops were set up. Factories that turned out handles and household woodenware were also established. Thousands of top-quality trees were taken out for use as ship timbers. Enormous quantities of wood were consumed in tile, brick, and lime kilns. Railroads bought and used much native timber as ties and bridge timbers. The opening of oil fields created an enormous demand for wooden "sucker rods" to be used in pumping oil. A vast amount of good timber went into rails and posts for fences.

In Paulding County, adjoining our county at the north, big charcoal ovens were built adjacent to a furnace for reducing iron ore, which was brought down from the Lake Superior region. This establishment yearly consumed the timber from about one thousand acres, converting some one hundred twenty cords of wood per day into charcoal sufficient for making forty-five tons of iron.

A tract of twenty-five acres or more of timber was commonly left when a farm was cleared. This was intended to provide fuel and lumber as needed in later years. The woods of course provided shelter for game birds and animals, so that for years there was excellent hunting. In many cases tracts from which virgin timber had been removed and sold were left undisturbed until small trees and saplings developed, producing a new generation that was known as second-growth timber.

Whites learned early from the Indians the art of making maple syrup and sugar. For a long time these maple products were the only sweets other than honey available in pioneer settlements. When they cleared their land, many spared large groves of maples. Sugar or hard maples were preferred, but much sap came from soft maples. On those farms sugar-making was long a regular spring occupation.

Living in our community was an old man understood to be part Indian. Each year he cultivated a patch of tobacco. After curing, he reserved some of the leaves for smoking. He prepared others for chewing by boring large holes in maple trees and packing the tobacco into them. The sap gave the plug qualities and flavor that he considered highly desirable.

Among native forest trees were beeches, hickories, oaks, sycamores, walnuts, elms, maples, wild cherries, mulberries, ashes, and box elders. There were also honey locusts, coffee nuts, and ironwoods. Native poplars were known as "cottonwoods," basswoods were commonly called "linns." Shrubs included sumac, dogwood, elderberry, pawpaw, spice-bark, haws, buttonwood, prickly ash, and several varieties of wild berries. Wild grape vines with thick, heavy stalks grew to the tops of some of the tallest trees. No evergreens of any kind grew native anywhere within many miles. A neighbor once found in the woods a small red cedar. Transplanted near his home, it developed into an attractive tree. No doubt that cedar grew from a seed carried by a bird from a tree growing many miles to the northeast where red cedars are common. Although the buckeye, or American horse-chestnut, flourishes over much of Ohio, giving it the nickname "Buckeye State," none grew on our farm or anywhere near. The redbud, or Judas tree, common in woodlands and beside streams in nearby areas, was also conspicuous by its absence.

In early spring the ground in woodlands was carpeted with wild flowers. Predominent among them were violets, sweet Williams, buttercups, Dutchman's breeches, Jack-in-the-pulpits, deer-tongue lilies, and May apples. Few ferns were found. The Virginia creeper was also rare. We children often gathered large bouquets, being partial to sweet Williams, wild roses, and blossoms of the wild grape, because of their agreeable perfume.

For many years all fields in the region were enclosed by rail fences, laid up in the familiar zigzag or "worm" form. Weathering and decay made steady inroads upon the rails, many of which on our farm had seen half a century or more of use. Gradually, fences of this type disappeared from the scene, nearly all being eventually replaced by factory-made wire fences. Broken and discarded rails were bucked up and used as fuel. All odds and ends of other wood no longer useful met the same fate.

Some farmers bought Osage orange seedlings and planted them in hedges to replace rail fences. The wood of this tree, named for the Osage Mountain region and the worthless orange-shaped fruit it bears, was used by Osage Indians and others for making bows; hence it is sometimes called bowwood. The hedges added a picturesque touch to landscapes, but they were not satisfactory substitutes for good fences. Most of them have now been torn out because, after years of growth, they occupied too much space, and they robbed field crops of nutriment and moisture.

THREE

Drainage

CLEARING THE LAND OF TREES,
though on a limited scale in the earlier years of settlement,
aided in disposing of surface water. It allowed the sun to
dry the soil and to hasten the melting of winter snows,
which before had fed water into the ground during a long
period in spring. Drainage work began with the removal of
obstructions in natural streams and the digging of open
ditches on an extensive local scale. Later, drainage systems
became county and intercounty projects carried on under
regulations prescribed by state laws.

A notable drainage project was the "Jackson cut-off" in
Wood County. Completed in 1879 at a cost of $110,000,
this channel, nine miles in length and twenty feet deep in
places, connects headwaters of the Portage River to Bea-
ver Creek, a tributary of the Maumee. It has converted
into a garden spot 30,000 acres of land, formerly worthless
because the Portage River, meandering northeasterly to
Lake Erie, did not provide adequate drainage. Another
main-line channel, dug some years earlier in one of the
wettest portions of the Black Swamp, is more than thirty

miles long. Canal-like in proportions, it drains some fifty thousand acres.

Drainage work continued until the region was covered by a network of open ditches having an aggregate length of many hundreds of miles. The use of horse-drawn dump scrapers for scooping out these ditches involved an enormous amount of very strenuous human labor. In many cases the drainage channels follow the routes of old, shallow, natural watercourses, winding for miles through fields and woodlands of several farms. Some run alongside highways, being so routed to avoid cutting up and wasting farm land. After the automobile came into general use, miles of guard rails had to be erected beside deep roadside ditches. Where the danger to motorists was greatest, sewer pipes were laid in and the ditches were filled up.

These man-made ditches and many of the natural creeks and rivers have been deepened and widened at intervals since the first drainage work was done. In some cases this has involved, in the lower courses, blasting and removing long ledges of bedrock. The work, now done with mechanized equipment, goes on year after year. Wherever one may travel in the region, he is pretty sure to find a power shovel or dredge at work, increasing the water-carrying capacity of some ditch, creek, or river.

In the decade 1835–1845 several hundred miles of canals were dug and made operative in the state. The Miami and Erie Canal, completed in 1845, extended from Cincinnati, on the Ohio River, to Toledo, on Lake Erie, a distance of 250 miles. Its course lay over a part of the boundary line between our county and adjoining Allen County on the east. Together with a branch line to the north of us, extending from Indiana, it aided greatly in the drainage of the

22

area, although of course, like others of the system, it was intended to serve primarily as a waterway for trade and travel. In its heyday the Miami and Erie carried a great volume of freight, and hundreds of passengers traveled in its packet boats. The canal and the area it served soon came to enjoy extraordinary prosperity. This prosperity continued only a relatively short time, however; it was slowly eroded, and finally destroyed, by the competition of the railroads, the first of which in the state was completed in 1848—about the time the canal system was getting into its stride.

In the flourishing days of area canals, towns along their courses grew quite rapidly. At several points new towns sprang up suddenly and burgeoned forth phenomenally. Without the canals they probably would never have come into being at all.

Some of the canal workers and a considerable proportion of the people attracted to canal towns were rough, lawless adventurers. Saloons for the accommodation of roisterers and gamblers in many cases predominated among the business establishments of such towns. One canal town that now can boast only about fourteen hundred residents had a population close to five thousand and some forty saloons in boom days. Behind the bar in saloons that catered to canal trade, proprietors in numerous cases maintained a sort of stockade. When carousing boatmen or other patrons got drunk and began to brawl, a muscular bouncer hustled them into the "bullpen" and left them there to fight it out.

Sedimentation and the trampling of livestock cause creeks and open ditches to choke up fairly rapidly. Grasses and weeds help the process along. The creek through our farm, like most others, had to be dredged out at intervals of five or six years, the work being all done with horse-drawn scrapers. The gumbo, mixed with a conglomeration of weeds, grasses, sticks and stones, was scooped out and dumped in uneven piles on the banks.

Wherever a fence crossed a stream, it was necessary to maintain a floodgate to prevent livestock from escaping under the fence. Built of boards to fit the contour of the stream bed, it was suspended horizontally and hinged so that it could swing downstream when the water rose, then close automatically as the water went down. We always had to take down and later remount our floodgate—sometimes to rebuild it completely—when dredging was done.

Open ditches carried off surface water, but they did not help greatly with water in the soil or with water that stood in distant low spots. To remedy this condition, early farmers laid box-like plank drains in their fields, designed to discharge into open ditches or creeks. They helped a great deal, but, though made of good oak, they never lasted long because they were subject to conditions highly favorable to decay. The channels they had provided gradually filled up with soil so that at last little or no water could get through.

In time burned clay tiles, generally handmade, came into use for underground drains. Having round tops and flat bottoms, they were known as "horseshoe tiles." Few of the drains in which they were used were satisfactory because the tiles were easily broken down and because outlets generally were so shallow that they could not be laid deeply enough. Often the plow, running at a depth of seven or eight inches on our farm, would crash through old horseshoe tiles of drains put in by our predecessors.

24

Though choked with dark loam, we usually found them still carrying a trickle of water. We also found remains of old plank drains. The wood was so badly decayed that only fragments of the original boards remained. The black soil that filled the crumbling ducts, however, was so porous that some water was still getting through it.

Those inadequate old drains stood pathetically as evidence of human courage and fighting spirit in the face of grimly adverse conditions. The original owner was aware that outlets were not deep enough for those drains to be fully effective. But they were the best available at the time, and he was willing to invest time, labor, and hard-earned money to make the most of resources he had.

After an extensive system of deeper outlet channels had been provided, mills for making good cylindrical clay tiles were set up throughout the area. These plants, not greatly different from those now in general operation, had machines for grinding and pugging the moistened clay, then extruding the tough plastic material through a die in the form of a continuous tube. Diameters ranged up to twelve inches. As the tube came from the die, it was cut by a tautly stretched wire into standard twelve-inch lengths. After drying in open air under a roof, the tiles were burned in wood-fired kilns until hard. The demand for clay tiles remains high. Many, of cast concrete, are also made and sold. Even though his fields may be underlain by numerous lines of tile drains, the average farmer continues to plan for more.

The underground drainage system of the old Black Swamp is now probably the most extensive in the world, connected into innumerable open ditches and natural streams. Even heavy rains are disposed of quickly, the runoff being hastened by widespread deforestation. At times enormous volumes of water, augmented by melting snows, pour down and cause serious flooding in lower outlet channels. Good soil, totaling millions of tons, is carried away, being finally deposited as navigation-obstructing silt in Maumee Bay.

The soil water table has been gradually falling for years, at a rate approximately commensurate with the progress of deforestation and the rate at which deep drainage has been extended. With the water table too low, growing crops suffer in periods of scant rainfall. Since about 1935, farm and municipal wells here and there have been failing in dry summers. If deeper drilling does not restore water supplies, it becomes necessary for farmers to buy water and for towns to impose use restrictions. At great expense some municipalities have built reservoirs for storing water pumped from dammed streams. Others are drawing water through pipe lines from distant sources. In a few instances cities having abundant supplies are selling treated water to neighboring municipalities and populous rural areas, metering it through large mains.

For years, all trenches for tile drains were cut laboriously with spade and shovel. Now, practically all trenching is done by a self-propelled machine that does the work a great deal faster and saves an enormous amount of time and labor. However the trench may be cut, tiles are placed in the bottom so that each fits squarely against the one previously laid. Joints are never cemented or otherwise sealed; they must be open enough to allow water to seep in, yet no joint opening must be so wide that soil will be washed in, to be lost or to accumulate in the drain.

26

A pair of knee-length rubber boots was a must for the ditcher working with hand tools in sticky mud or in stretches of trench containing water. The soles were thick and tough, an important advantage in spading and in kicking tiles into place. In the course of about a decade, during which I had a hand at putting in an aggregate of several miles of tile drains on our farm, I became pretty well acquainted with those boots. They were not only heavy to drag about but they had a provoking habit of working the wearer's socks halfway off, so that rolls and wrinkles in them cramped and pinched his feet. They were too warm in warm weather, and feet nearly froze in them when it was cold. Walking through a deep trench, the flaring tops funneled in a lot of loose earth, which made the feet uncomfortable and soiled the inside of the boots and, thereafter, socks and feet.

Water flowing from a tile drain is nearly always as clear and sparkling as water from a spring. It is pleasantly cool, even in summer—very tempting when one was thirsty. Some drank it freely, but this was risky because of the possibility that it might be contaminated.

FOUR

The Soil

THE TOPSOIL OF OUR FARM
and others about it was a dark loam averaging about
twelve inches in depth. Small, widely separated patches
that lay a little higher than surrounding areas were found
in some fields. Farmers called such patches "clay knobs."
The soil here, dark gray in color, with a texture quite
different from that of the other soil, was much less fertile.
Water did not readily penetrate it, and it became hard
when baked in the sun. No doubt the knobs were weath-
ered hillocks of glacial clay that, by reason of their original
relative height, were never enriched by the deposit of
sedimentary material as were surrounding areas. Beneath
both the knobs and the dark soil lay a bed of yellow clay
several feet thick. Underlying this was a much thicker
layer of tough, bluish clay, highly impervious to water.

In dry summer periods wide, deep cracks opened in
the dark soil, attesting its marshy origin. If fields were
worked when a little wet in spring, big clods were formed
that might last through the summer. This tendency of the
soil to pack and form clods gave rise to bothersome farming problems. Since the condition was common, farmers
often alluded to themselves as "clod-hoppers." Nowadays,

cultipackers and other improved implements take care of the clods and make field work much easier. A slightly larger proportion of sand in the soil would have improved it greatly. This was demonstrated by a man who worked several tons of pulverized limestone into the gummy soil of his garden: the stone not only solved the clod problem, but it made the ground much easier to work.

Our fields were capable of growing good crops of corn, oats, wheat, rye, potatoes, and grasses for hay and pasturage. Some about us occasionally sowed a field to barley, but the barbed, scratchy beards that anchored themselves in clothing at threshing time tended to lower the popularity rating of the grower. Soybeans, sugar beets, tomatoes, and alfalfa, though now familiar and important crops in the region, were not grown in my time on the farm. Sweet clover grew wild in patches about us. Called "bee-weed" by some, it was commonly regarded as an ill-smelling nuisance. Some years later, farmers began planting it extensively as a soil builder and a forage crop.

Occasionally, as we plowed our fields, boulders were brought to light. Most of the stones were so small that they caused little trouble. Now and then a big one would throw the plow out of the ground and damage its point. Such stones were dug out and hauled to some out-of-the-way place where they would be a bother no more. The largest boulders were about the size of a bushel basket. Bigger ones might be found if we dug much below plow depth. Most field stones were granite, some in varied shades of red, some blue or greenish in color. All had been glacier-borne, in all probability, from the Pre-Cambrian granite surface of the Canadian Shield (known also as the Laurentian Shield or the Pre-Cambrian Shield), a vast area lying north, west, and south of Hudson Bay.

Big boulders were not always dug out and hauled away. Men sometimes dug deep holes beside them, then rolled them in and covered them, well below the depth at which plows ordinarily run. Some ridiculed this practice, declaring that the erratics would work upward in time and again become obstructions at plowing time. While this argument appears to ignore completely the law of gravitation, there was evidence that repeated freezing and thawing, which favored the deposit little by little of additional soil beneath the stones, actually lifted them gradually. I have seen boulders reappear in fields previously made free from plow-depth stones by burying some and removing others.

We picked up in our fields a few arrowheads and other stone artifacts left by Indian hunters and warriors. Like field stones, these were less common on our farm than on the slightly more elevated lands of neighbors. This was probably because the latter were originally relatively drier than surrounding areas, and for that reason conditions there were more favorable for hunting.

Because of peculiar soil qualities weather conditions have always been of special importance to Black Swamp farmers. Given reasonably favorable weather—other things being equal—we usually saw the competent farmer come through the year with a favorable balance. The wrong kind of weather, though it might prevail only during a short period, could seriously affect profits.

Unfavorable weather was not the only thing that could cause trouble. Livestock diseases or insect infestations might cut deeply into incomes. Cutworms at times worked havoc by eating off tender shoots of corn. Army worms appeared some years to prey upon crops. Grasshoppers chewed away at field plants throughout every summer, but they never came in such numbers as to cause widespread damage.

Probably no insect pest caused as much crop damage as the chinch bug, which, coming originally from Central America, quickly spread to sections of North America in which grain crops are grown. Almost every year great hordes of these evil-smelling insects would appear in wheat fields and suck from the jointed stalks sap needed to mature plump, sound grains. After the wheat was harvested, they lost no time in moving into the oats. After oats harvest, the pests, like an invading army numbering millions, would make a beeline for the corn, entrench themselves on the stalks, and proceed, vampirelike, to suck the very life blood from the plants.

Farmers generally felt themselves helpless against the determined bugs. A few cut shallow trenches between fields and poured into them sticky creosote tar to prevent migration to untouched foraging grounds. This was laborious and expensive, but it could save crops from damage.

FIVE

Above and Beyond the Call of Duty

SUBDUING THE WILDERNESS,
establishing homes, and conditioning the land to enhance
its productive capacity involved an enormous amount of
wearing toil. The pioneer men of course did the arduous,
backbreaking work, but it is in the record that the valiant
women of that day bore an astounding share of the
burden. Few today have any true conception of the scope
and volume of the work that women on farms did until
a generation or two ago; probably, not many would believe
the story if it were related in full. There was then a great
deal more truth than poetry in the couplet "Man works
from rise to set of sun, but woman's work is never done."

With only slight diminution, resulting from rather limited
improvements in living conditions, farm women continued,
within the memory of many now living, to follow the long
established tradition of performing numerous tasks in addition to those involving ordinary housekeeping and the
care of children. Naturally, this imposed upon some bur-
dens too great for their strength. Monuments in old cemeteries bear the names of many women who succumbed in
what are now regarded as the middle years of life. Un-

questionably, overwork was responsible in many cases; but diseases took a heavy toll—men, too, died at a much earlier age than in later times. Complications incident to child-birth of course also resulted in a considerable number of untimely deaths.

In every farm home there were the familiar, unending housekeeping tasks. All involved real work, all were time-consuming, and about all had to be done with equipment that to moderns would seem very primitive. Generally, there was a sewing machine. Nearly every housewife had some kind of hand-operated washing machine with an attached wringer. But in most instances all water had to be pumped by hand and carried in pails. One of us boys was nearly always assigned to help Mother with the heavy tasks of washday. There was no way to heat water except in a wash boiler on a wood-burning stove, whose heat in summer might cause much discomfort. Sadirons had to be heated on the same stove on ironing day.

On farms about us we occasionally saw women working with their men in the fields. Often, they seemed to welcome field activity as a diversion from housework. On a few farms the women did all of the milking—often there were six or more cows. Whoever did the milking, the women attended to straining and storing the milk in crocks twice daily. They hand-skimmed it all with a dipper every day to collect the cream.

The job of churning came two or three times a week. Some used a tall woodstave churn, but the majority seemed to prefer one of earthenware. In either case there was a wooden dasher on a slender vertical handle that had to be plunged up and down, sometimes for an hour, until the butter "came." Then the fatty particles were dipped out and worked with a wooden paddle in a wooden bowl to remove liquids. Finally, the mass was salted and shaped into rolls.

After using her dasher churn a long time, Mother got one of the barrel type, turned by a hand crank. This made the job a little easier and faster. About the same time, she bought a gravity-type separator. In this, whole milk was stored in a water-cooled tank. After the cream had all risen to the top, the milk beneath it was drawn off through a spigot at the bottom. Soon after this, owners of large herds of cows began using hand-cranked centrifugal separators. Milking machines did not appear until years later, when electricity became available.

The over-all management and care of poultry were nearly always left to the women. Yearly, each raised an average of about one hundred young chickens; often, the same number of turkeys. Some also kept a few geese or ducks. They entrusted all hatching eggs to chicken hens for incubation. When the baby birds appeared, they fed and cared for them regularly. The hens that had done the hatching gladly took on the task of brooding and protecting the young until they were able to make their own way. For a long time there were no incubators, brooders, or commercial hatcheries. The culling that was done, generally in the fall, involved only weeding out birds that were inferior in size or robustness or that were considered to have reached retirement age. Commonly, a dozen or more healthy, old hens among the culls were reserved for cooking through the winter; the remainder were sold.

Men of the family did all of the rough, heavy work. Children at times gathered the eggs and did some of the feeding; the women, however, supervised all of this work

37

and did much of it themselves. Nearly all of the money for their personal use came from the sale of poultry, eggs, and butter. Often, part of this produce was traded for needed supplies of groceries, but cash balances received in the course of a year, though known as "pin money," added up to a sum that was by no means negligible.

On most farms, women assumed principal responsibility for the vegetable garden after the ground had been plowed and the rougher preliminary work had been done. They collected, dried, and stored seeds and grew their own transplants, starting them early from seeds indoors. Seeds that could not be grown conveniently in the home garden were ordered from the catalogues of big seed houses. They drafted a good deal of help, supervising the work and doing much of it themselves.

As soon as frost was out of the ground in early spring, Mother began bringing in for table use parsnips grown in the garden the previous summer—it was commonly believed that these vegetables were unfit for use until they had gone through the winter frozen in the ground where they grew. At the same time freshly grated horseradish, from roots that grew in garden clumps, would appear upon the table. Soon afterward came tender rhubarb shoots to be stewed or baked in pies; rhubarb was an infallible harbinger of spring.

Bushels of tomatoes and tree-grown fruits were canned. Numerous kinds of pickles and relishes were made and stored. Quantities of sweet corn, peaches, and apples were spread out in the sun to dry day after day, protected from insects by mosquito netting. A quantity of sauerkraut was turned out each fall. This tended to be a family project, some cutting and chopping cabbages while a muscular individual, with a big wooden "stomper," pounded the shredded and salted vegetable in a big earthen jar until it became a pulpy, juicy mass. In early winter about a bushel of shelled corn was converted into hominy. In the process the grains were treated with lye from wood ashes, then stirred and rubbed with a wooden paddle until the tough skins were removed from the kernels. Finally, all traces of lye were removed by thorough soaking and washing.

Most housewives did a great deal of work at butchering time in connection with sausage-making and the preservation and storage of meat products. As matters of course, they tackled the tasks of soap-making in spring and turning out a big batch of apple butter in the fall. Much time and effort were involved in these projects, for the best equipment they had was primitive and clumsy.

For soap, waste fats were collected throughout the year. Wood ashes from stoves were stored in barrels, protected from rain. When mild spring weather came, water was poured into the barrels, blocked up on wide boards so arranged that lye from the ashes would run into crocks beneath the boards. Lye was poured over the grease in a large iron kettle. The mixture was boiled gently and stirred from time to time until saponified. The finished product, of about the consistency of vaseline, was used for scrubbing, laundry work, and dish washing. Commercial products for cleaning and scouring were limited both in quantity and quality. Many housekeepers used brick dust for rough scouring. They prepared it by crushing and powdering bricks with a hammer.

According to a family story, my father's grandmother, wife of an early settler on the Ridge, was one day making soap over an outdoor fire when a group of Indians stopped at the cabin to beg something to eat. Approaching the kettle, they noted its bubbling brown contents. With happy cries of "'Lasses! 'Lasses!" each cautiously dipped in a finger, then licked the finger. The taste was such a profound surprise that a dour, angry look came over their faces. A moment later, apparently struck by the humor of the situation, their expressions changed and they stalked haughtily away.

Many farm women grew flowers in big beds about the house and in the lawn, giving countless hours to sowing seeds, transplanting, weeding, and watering. They took considerable pleasure in showing off their finest blossoms to friends and in swapping seeds, bulbs, and cuttings. When fall came they moved their finest specimens into the house, placing them in pots on special shelves and on tables before sunny windows. Throughout the winter they tended them with loving care, reveling in the color and fragrance of the blossoms.

With the exception of their best coats and dresses, many women made almost all of their own clothing. They also made most of the clothing the children wore. Somehow, they found time to keep the youngsters' garments mended —in spite of rough wear, their clothes were nearly always clean and in good repair. It was a common practice in farm homes to buy yards of unbleached linen toweling and cut and hem the material into suitable lengths to replenish the family towel supply. The new material was stiff, and woody fibers entwined in the woven threads made it a little harsh. The towels improved with use, however, in time becoming snow white and quite soft. Washcloths, for those not addicted to the use of a natural sponge, were ordinarily cut from sound parts of discarded cotton under-wear.

In farm homes much time was given to making quilts and comforters with interlinings of cotton or wool batting. Outer faces were usually "pieced" from scraps of cloth to form various designs. The best parts of the family's worn-out or outgrown clothing went eventually into rugs and carpets. At our house Mother dyed much of the cloth for this use. Unlike women of earlier generations, who had to go to the woods and swamps for dyestuffs, she enjoyed the advantage of commercial dyes. The material was torn or cut into strips that were sewed end to end and wound into large balls.

In this connection comes to mind one of my very earliest recollections. My maternal grandmother, a guest in our home, busied herself one day at sewing carpet rags. After watching the work with the fascinated interest of a .small child, I expressed the desire to do some sewing too. There was no blunt refusal, no argument, no effort to shunt my interest to something else. Grandma simply tied a thread to the head of a pin and passed it to me. I went blissfully to work, "sewing" away in perfect content. In more mature years this has impressed me as a classic example of prac-tical psychology applied to dealing with a child's harmless whim or fancy.

Mother took her carpet-rag balls to a neighbor who had a carpet loom. The sewed rags formed the woof of the fabric; commercial cotton yarn, known as "carpet chain," was used for the warp. Nearly all of our carpets and rugs were hand-loomed from these materials. Ordinarily, the size of the loom limited the width of the fabric to about thirty-six inches; strips of the required length were cut from the roll as it came from the weaver and sewed edge to edge to form carpets. Clean cloth remnants not suitable for piecing quilts or for use as carpet rags went into a rag bag to be finally sold to a ragpicker.

Few women ever shirked any phase of the work involved in seasonal housecleaning. They went over all rooms from top to bottom, dusting, scrubbing, scouring, washing, and polishing. At our house we boys helped with the carpets (all rooms except the kitchen had carpets). We pulled out the tacks that held them in place, carried them outdoors, and beat them. We took out the old straw that underlaid all but the best "store" carpets, then swept and dusted the floors. After the floor boards had been scrubbed, we brought in fresh, clean straw, spread it evenly, and then relaid the carpets.

For luxurious comfort in winter everybody had feather-beds—ticks filled with choice duck or goose feathers. Soft and warm, they took much of the discomfort out of sleeping in icy bedrooms. In severely cold weather some cold-blooded people slept between two feather-filled ticks. Collecting and processing the feathers were tasks that fell to the lot of the housewife. Many of the beds were passed along as heirlooms from mothers to daughters. All those in regular use had to be exposed to air and sunlight at times throughout winter and spring, no trivial chore for the housekeeper.

One summer a man driving a van went from door to door soliciting business as a "feather renovator." Contents of featherbeds entrusted to him, he explained, would be run through a machine in which live steam would clean and sterilize the feathers. They would finally be returned in their owners' ticks, he promised, thoroughly clean and

42

extraordinarily fluffy. No complaints were heard from any of the fellow's numerous customers. It is possible, however, that all had been roundly cheated, for in some cases the business appears to have been operated as a thieving racket—the operator stole and marketed several pounds of good feathers from each tick. The customer would suspect nothing because the ticks, filled with fluffy feathers, came back much plumper and firmer than when they were taken away.

When summer came, ticks filled with straw were substituted for featherbeds. Soon after the threshing was done, the crushed, broken straw was discarded, and the washed ticks were stuffed with new, clean straw. One fall as we shucked the corn, we collected clean husks to be used instead of straw. This turned out to be a long, tedious job, for each tick required a lot of husks, and several had to be filled. The husks wore well; they didn't crush and break as straw always did.

Whether for feathers, straw, or other filling, ticks were simply flat, bed-size bags made of ticking, without throughties or other fastenings for holding contents as a flat pad of uniform thickness. Therefore, bedmaking was likely to involve at times considerable effort to smooth out and level wads and lumps that had formed.

Housewives baked bread and rolls once or twice each week; baker's bread rarely appeared on any farm table. Like most of her neighbors, Mother made her own yeast. Each summer she gathered hops from vines in the garden and dried and stored enough for a year of baking. About twice a week she baked a batch of half a dozen or more pies. Cake was available for at least one meal each day; cookies and doughnuts, or crullers, were always at hand.

There were scores of top-flight cooks among the women. Quite naturally, no doubt, I have always believed that my mother stood near the top of the list. Some of the most expert seemed rarely to follow any written recipe; it appeared that an unfailing instinct told them just the right quantity of each ingredient to use, just the proper time to add it, and just how much heat and how much time were required for perfect final results. With no reservations whatever, competent judges would have awarded blue ribbons to some of the meals, prepared without benefit of any but the simplest of kitchen equipment.

"Company" at our house and in a good many other farm homes was almost commonplace. Company meant deluxe meals, with ham or chicken—often both—and a bewildering array of vegetables, fruits, relishes, and desserts. Sunday family dinners were nearly always special meals, with an unusual variety of foods in unstinted quantities. All this, of course, required an enormous amount of time and effort.

Girls in farm homes were able and willing assistants at most of the tasks their mothers had to do. If there were no daughters, the farmer's wife generally had to depend on a hired girl or go it alone. It seemed rarely to occur to males of the average farm family that they could or should lend a hand.

SIX

The Well-fed Farm Family

THREE SQUARE MEALS A DAY
is an expression that must have had its origin on a farm.
Anyway, that is what most farm families had—three meals
daily, all of them "square." The average family ate meat
of some kind at almost every meal, flanked by generous
portions of other substantial foods. Town people ate almost
as heartily. They seemed to feel that no meal, even breakfast, amounted to much unless it included big steaks or
bacon with potatoes or equally nourishing viands. Yet the
proportion of fatties, urban and rural, was much smaller
than in later times, probably because much more physical
activity was then required, so that big food intakes were
converted into muscular energy instead of going to form
gobs of unwanted fat.

Farm families' bread was practically all home-baked, in
the ovens of kitchen stoves—only one outdoor "Dutch
oven" was known in the area. Unlike bread from modern
commercial bakeries, loaves that came from home ovens,
of course, never contained dough conditioners, special
vitamin "enrichment," sodium propionate, or other strange
chemicals. A baking might number eight or more big
loaves. If the last of a big baking became a little dry,

it might be toasted by placing slices on a hot stove lid or by holding them over glowing coals. Some dry bread was eaten as French toast, but it was generally called "fried bread." At our house dry loaves were sometimes placed in a deep utensil above an inverted perforated pan and exposed to steam from boiling water. This made the bread almost as good as when it first came from the oven.

Flour came from home-grown wheat. Ordinarily, we took about four bushels to the mill at one time. The miller, of course, took toll from our wheat to pay for his services, so that we would get back about two bushels of white bolted flour, maybe half a bushel of bran, and a like quantity of "middlings" or "shorts." The latter included principally the seed germs from the grains, composed of oils, proteins, and mineral elements, actually the richest and most valuable constituents of the wheat. In this by-product were also stores of vitamins, although at that time these elements were unknown and unsuspected. We used the bran and middlings, with some that we bought, as supplementary feeds for livestock.

Often, we boys ate some of the middlings ourselves. They were pleasantly chewy and agreeable to the taste unless they had become rancid, as they would with age. We liked to eat wheat, too. It had a pleasant taste, and when well masticated, became a waxy mass that could be chewed like gum. As a matter of fact, it was about equal to one kind of chewing gum then sold in stores. This gum, made of paraffine wax, sweetened and flavored, came in the form of round bars like sticks of candy. One had to grind on it steadily or it would crumble and slip down the throat. Chicle chewing gum was also offered in stores. It was much to be preferred, but, as is so often the case,

48

when one bought it he had to sacrifice much in quantity to get quality. Occasionally, we peeled long strips of the mucilaginous inner bark from slippery elm trees to chew. It had an agreeable flavor, and a cud of it would last for hours. At times we carried in our pockets rolls of the stuff big enough to keep us grinding several days.

One winter a family in the neighborhood bought some spring-wheat flour produced by a big Minneapolis miller. They were so loud and so enthusiastic in their praise of the superior quality bread it yielded that within a few weeks dozens of farmers thereabouts were swapping hard winter wheat from their bins for spring-wheat flour. It did make good bread, and some housewives declared it was easier to handle on baking day. Though more costly than flour from home-grown wheat, it is questionable whether it could have been proved to be superior in any respect.

Corn from farmers' own fields also contributed much to their larders. It was by no means as important as a food item in our day, however, as it was in earlier times; for the pioneers it was a mainstay. Growing their corn involved some difficulties, but their real problem was getting it ground into meal—grist mills were few and far between. We were told that the first owner of our farm had to take his corn for grinding to Piqua, some fifty miles distant.

One of the first families to establish themselves in our county was that of William Miller, one of my father's forebears. The mill nearest his clearing on the Ridge was at Willshire, the first organized town in the county. He had to allow two days for the 25-mile round trip because the roads were very poor and he always carried several bags of corn on the back of his horse—it was his custom to provide enough meal to feed his family and have a surplus on which he could draw for the benefit of Indian friends.

Setting out for the mill one morning, he requested one of these red men to look to the safety of his family until his return. That night nine stalwart braves strode into the Miller cabin, wrapped themselves in blankets and, without a word, stretched out on the floor. Mrs. Miller, knowing nothing of her husband's arrangement, was puzzled and more than a little frightened. To be on the safe side, she took up a position beside the fireplace, within reach of the big poker, and sat awake all night, tensely watching for any hostile move. At dawn the Indians, as one man, arose and marched silently from the cabin.

As soon as our corn matured in the fall, we dried and shelled select ears and had the grain ground at the mill. Meal from new corn has a quality and flavor that are never equaled by meal from old grain. One year we had a surplus of giant yellow popcorn ground into meal. It was superior to ordinary meal, with a deeper yellow color and a more pronounced corn flavor. Prime roasting ears from the cornfield (generally called "hog corn") were frequently served in lieu of sweet corn. Homemade hominy and cracked corn or coarse grits from the store often appeared on farmers' tables. Rice was also a favored food item; cold boiled rice, with cream, nutmeg, and sugar was eaten at times as a dessert. Cooked rolled oats, served with sugar and cream, was a substantial part of many farm breakfasts.

Pancakes made from wheat or buckwheat flour or corn meal figured prominently in many a farm breakfast. Soda biscuits also had a big following. These were all prepared in the home kitchen as needed, almost wholly from farm-produced ingredients—no one then dreamed of commercially prepared ready mixes such as are now widely used. Pancakes and hot biscuits, well buttered, were commonly eaten with sausages. Nearly always, there was also a generous supply of syrup. This might be sorghum, hot brown-sugar syrup, or a commercial product from a gallon container.

Farm families used apples in a big way. In winter Dad took ours as needed from the pit in which we had buried them, chopping through the frozen earth to get at them. They were always crisp and firm, with a slight taste and odor of the straw that had covered them. Other fruits were used freely, fresh in the summer, canned or dried in winter. In addition to garden-grown berries we occasionally picked some that grew wild, including gooseberries that, porcupine-like, bristled with sharp spines. Preliminary to their use, it was necessary to shake them vigorously in a covered pail, to break and flatten the stickers. Rhubarb, commonly called "pie-plant," grew in every garden, and was widely used in pies and desserts. By way of variety, some at times stewed it with a few cherry leaves, which gave it a cherry flavor.

We ate melons at any convenient time, often right in the patch where they grew. Occasionally, soon after the first frost, we stored the best remaining watermelons inside a shock of corn. They kept in perfect condition for weeks, being especially enjoyable because they were chilled exactly right.

Almost every grocer kept a bunch of bananas hanging within convenient reach. They were sold by the dozen rather than by the pound, as in later times. Farm families did not buy them regularly, but they nearly always appeared at special big dinners. Oranges were consumed in quantity only at the Christmas season; they were regarded as special treats. Coconuts, Brazil nuts, filberts, almonds, English walnuts, and pecans contributed to both Thanksgiving and Christmas feasting. Lured by the aroma that came from the roasters, farm people in town, at circuses, and at the fair ate many bags of peanuts, sold at what would now be regarded as extraordinary bargain prices. One could buy hot chestnuts in season at candy stores and fruit stands; none grew locally.

Generally, home-grown peaches were of inferior quality because they came from unbudded seedling trees. Riding in a train one time, Dad bought a few extra large peaches from the news "butcher." He liked them so well that he brought home two of the seeds and planted them. Shoots that came from them grew vigorously and within a few years began to bear large, juicy peaches of excellent quality. This was a very unusual outcome, one that could be expected in not more than one case out of a hundred; the only way to be sure of quality peaches is to plant trees propagated by budding from stock known to be good.

It seemed to us boys that those who considered the pawpaw a delightful palate treat must have had a strangely perverted sense of taste. Year after year we tried some of the soft yellow fruit but were unable to enjoy a single bite —to us the taste was anything but agreeable. Our verdict was the exact opposite in the case of wild mushrooms. For us it was a most happy occasion when, mainly by accident, we came upon specimens of the sponge variety, the only one we knew to be non-poisonous.

Oysters, which came in flat gallon and half-gallon cans, appeared at times on farmers' tables, especially at late evening suppers and parties. About the only other seafood eaten was whitefish, packed in salt brine in wooden pails. The fish were soaked overnight in fresh water, then rolled in corn meal and fried, often for breakfast.

Eggs were eaten freely but no family could begin to keep up with the output. Some of the boys in school boasted of having eaten eighteen or twenty Easter eggs. We could believe this, but their stories of stealthily hiding eggs to get enough for this feast were taken with a large sprinkling of salt; no one, we felt sure, had to go to such lengths to enjoy all he could eat.

An enterprising store in a Welsh community some nine miles distant sent out a "huckster wagon" in summer. This was literally a store on wheels, carrying almost every grocery item and, in addition, such things as cotton yard goods, notions, brooms, and even kerosene. The spring-mounted wagon each day covered some part of the territory within a radius of thirty miles. Drawn by a span of wiry mules driven at a brisk trot, it reached each customer fairly regularly at a certain hour on the same day in each week. This saved as many a trip to town, since eggs could be sold or traded for needed supplies. As small children, we were always at hand to welcome our hucksters, knowing well that before he left, the shrewd fellow would present a stick of candy to each of us.

Milk was among the most abundant of food materials. It was used in about every known way, except that in many homes sweet milk went begging as a beverage— buttermilk was preferred. No dairy product except butter had much market value because a large part of the people in town kept cows. So, on the average farm, many gallons of good skimmed milk and buttermilk every day went to pigs and calves.

Commercial ice cream was eaten only at soda fountains and church socials. Milk and cream for commercial uses were carelessly handled. There was no pasteurization, no adequate refrigeration, little or no inspection anywhere; and ice used in freezers came from ponds unprotected against contamination. Consequently, some commercial ice cream was a menace to health. We youngsters made ice cream in winter by stirring sugar and flavoring extract into half-frozen milk. If the milk was not frozen, we added snow to the mixture. Often in summer, we made ice cream in a two-gallon freezer.

Some friends in school were loud in their praise of buttermilk pies. Eating such concoctions, we felt, would be the height of silliness. But we enjoyed pies that others doubtless would have regarded as strange and unorthodox. One was made from half-ripe grapes, another, from half-ripe tomatoes. We also liked pies filled with a mixture of sweetened vinegar, rich cream, and spices, thickened with cornstarch. This pie probably would horrify a dietitian, but we never worried about any dietetic defect or deficiency it may have had.

One day my brother and I caught and dressed a big possum. Against her better judgment, Mother roasted it for us to a perfect brown. We carved and ate, smacking our lips and swallowing with a great pretense of relishing the meat. Actually, each bite all but nauseated us; the animal was so fat that each morsel dripped with grease. We resolved that when we caught our next possum (fortunately, there never was a next) we would boil it and skim off the oil before it went into the oven.

A group of us one time dug a little pit, kindled a fire in it, and chucked in amid the smoking sticks several roasting

54

ears sheathed in their husks. Our cooking in this instance would have been voted a most dismal failure by any competent authority, but we shucked out the ears (only slightly warmed and strongly flavored by the kerosene used to start the fire) and ate them with gusto, all proclaiming that no corn was ever more delicious.

I once caught an exceptionally large turtle. Having read repeatedly that turtle soup is a surpassing epicurean delight, I determined then and there to put those glowing accounts to a practical test. Hatchet in hand, I put the big fellow on a block and teased him with a stick until he forgot his natural reserve and extended his neck beyond the shell. Quickly, I brought the hatchet down and severed his head. I set to work with hatchet, knife, and chisel to get that reptile meat out of the shell. Blood flowed freely. Soon everything, including my hands and the tools, was smeared with gore. My appetite for turtle soup waned rapidly as a very queasy feeling came over me. Within another minute, too sick to stand, I had to give up. I crawled away and lay half an hour under a shady tree. I then dug a deep hole and buried the bloody mess as quickly as possible.

The farm meat supply came from livestock grown on home acres. Cured pork was a basic element in the family fare during the greater part of the year. Butchering was a do-it-yourself project; generally it was on a co-operative basis, two or more neighbors—often brothers or other relatives—working together. Due to lack of refrigeration, the job had to be done after hot weather had ended. Ordinarily, one family killed a hog in October, the other family following suit a little later. Meat from each of these early butcherings, salted but not cured for long keeping, was shared by the participating families.

Hogs were commonly killed by a blow on the head from the poll of an ax. Some preferred to shoot them. Although the hapless pig was a familiar domestic animal, the man or boy who did the shooting was often affected to some extent by "buck fever." He did his best to center the bullet slightly above an imaginary line joining the eyes. If he failed to hit the right spot, he might try again, or the pig might be felled by an ax blow. Whichever way the animal was brought down, someone at once plunged a long knife into its throat and severed the big blood vessels there. Boys spoke of this as "cutting the jugular"; actually, both the jugular vein and the jugular artery, as well as other blood vessels, were opened.

Someone transferred boiling water from a kettle to the scalding barrel and threw in a handful or two of wood ashes. By this time all blood had trickled away. The pig was plunged into the hot water and two men soused it in and out, turning it end for end and round and round. If this work was all done right, the hair and outer layer of skin slipped off easily. Using sharp knives, it took only a few minutes to get the skin perfectly clean and white— even the blackest pigs were white when this job was finished. In case the scalding went wrong in some way, it was necessary to shave much of the surface. This gave rise to the expression, "a bad scald," often used in reference to any kind of undertaking that had not come out right.

The carcass was now suspended, head down. An incision was made, extending down the soft underbelly, and the mass of abdominal viscera, steaming in the cool air, was rolled out. Heart and lungs were removed from the chest cavity, and the carcass was washed clean. Cutting it into conventional pieces after it had cooled and stiffened completed the job.

The main pig-butchering came after real winter had set in. It was handled in about the same way, except that it was on a scale several time as large, to provide meat for the ensuing year. One or more uncles, with their families, came to help with our butchering; later, we helped them. Occasional tasks were assigned to the older boys, but on the whole it was a red-letter day of boisterous play and pranks for the youngsters who had been allowed to take a holiday from school.

After the cooled carcasses had been cut up and trimmed free of surplus fat, loins and other lean parts were cut into small strips or cubes and ground into sausage. This work, altogether occupying a full afternoon, was done indoors. Salt and black pepper were mixed into the ground meat— nothing else was ever added. Then, a generous portion was fried in a skillet on the heating stove. The aroma that came from that frying pan was something that no one could resist. Although we had eaten a hearty meal only two or three hours before, all hands fell to and ate big helpings by way of sampling the sausage to make sure that it had been seasoned properly.

The following day all fat parts, cut into small cubes, were cooked and put through a press that squeezed out the lard in the form of a thin, clear liquid. The residue of fatty tissues at the bottom of the press became round disks of closely compressed "cracklings." At times a few cracklings were baked in corn bread; some were fed to the chickens; the remainder were used for making soap.

Each family had its own pet recipe for curing hams, shoulders, and slabs of "side meat," as most folks called bacon. Soon after butchering day, Dad rubbed a carefully proportioned mixture of salt, black pepper, brown sugar, and a small amount of saltpeter into our meat. This was repeated several times, at intervals of a few days. We bought salt by the barrel, for this use and for the livestock. In the spring the meat was hung on rafters in the smokehouse and exposed to smoke from a fire in an old iron kettle on the earthen floor. Hickory wood was used for smoking, until we discovered that corncobs produced results quite as satisfactory.

Our home-cured bacon was never equal to good commercial bacon—it was too fat and not well streaked with lean. All who tasted the cured hams and shoulders, though, voted them superior on all counts to packers' products. Everyone considered the fresh sausage, hearts, and livers far and away better than any sold by butchers. To assure quality, we saw to it that all meat animals were bled thoroughly and at once after killing, that all were eviscerated as soon as possible, and that all internal parts retained for food were taken out promptly, separated from other parts, and cooled rapidly in open air.

Chickens were often killed for the table by wringing their necks. One seized the bird by the head and swung it round and round until the body pulled away and dropped to the ground. We boys were squeamish about this method, and always chopped off the heads with an ax. We felt that the ax did the job quicker and more painlessly, and blood seemed to drain away more quickly.

Every winter we had a quarter or a half of beef for our larder from one of our own animals or from one butchered by a neighbor. The meat might be preserved by allowing large pieces to freeze, then keeping them in an unheated space. Always, some was cooked, covered with hot grease, and sealed in containers. Occasionally, sizeable pieces were salted and dried, becoming practically equivalent to the old-timers' "jerky."

58

One winter we butchered a lamb. There was so much prejudice against the meat that we were a long time getting it eaten. Some of us felt, and declared unequivocally, that we "just couldn't go" mutton. A year or so later, Dad brought from the butcher a package of fresh meat that, when cooked, was eaten with keen enjoyment, all of us believing that it was exceptionally good beef. We sat shamefaced and silent when Dad disclosed, after the last morsel had disappeared, that we had just eaten mutton.

Some farmers now have their butchering done by professionals. Others buy dressed and packaged meats from custom butchers and store them in deepfreeze units. Much of the cured ham and bacon eaten at farm tables has been processed by commercial packing houses. Real homemade sausage and good home-cured hams are very hard to find.

SEVEN

Our Home on the Farm

SOON AFTER THEY WERE married in 1883, my parents, William and Nancy Jane Hook Good, established themselves on a Black Swamp farm of seventy acres. In all probability, Dad, then near the age of twenty-nine, was able to make a substantial down payment when they bought the farm. No doubt his father aided financially and helped to provide needed equipment; that was almost invariably the custom at that time and later. I never knew the price they paid per acre, but I am certain that it would appear trivial indeed compared with prices of later times (about 1955, similar land two miles distant sold for $700 per acre). Some thirty years after the initial purchase, an adjoining tract was acquired at approximately $100 per acre, making the total holding 140 acres.

Dad liked farming. That, I think, was a basic reason for his success as a farmer. He was an excellent manager, able usually to get work properly done at the proper time. He had most of the essential know-how, having had sound instruction from his father. He followed developments in agriculture as reported in leading farm journals, and he learned much by observing practices of good area farmers, of whom there were a goodly number.

My father's paternal grandfather, of Pennsylvania Dutch extraction, came with his family to Ohio from Lancaster County, Pennsylvania, in 1836. Born in 1854, my father lived to the age of ninety-two years. This farm was his home during more than sixty-two years. It was also Mother's home continuously from the time of her marriage until the end of her life, well past the age of eighty-six. Born in Greene County, Ohio, in 1858, she was a descendant of a Scotch-Irish Virginian who brought his family to Ohio in 1813.

They bought the farm from the family of Arch T. Priddy who had "entered" it—that is, bought it and obtained original title from the federal government—in 1839. At the time of my parents' purchase, a number of families thereabouts were living in frame houses. This farm, however, boasted only a log house; but it was something of a de luxe version of that type of dwelling.

Writing in a local newspaper on his ninetieth birthday in 1935, in reference to this house, a son of the original owner and builder recalled: " . . . There were no persons living . . . between Van Wert and his (Arch T. Priddy's) home. Farmers came from six miles away to help raise the house. . . . In preparation Father went into the woods to shoot some wild turkeys. . . . He heard a rustling in the leaves and spied a bear. . . . He shot the bear and bear meat, instead of wild turkey, furnished the main dish (for the raising). . . . One very cold night (some time after the house had been built) . . . an Indian . . . asked to come in and sleep. . . . A blanket was thrown on the floor . . . in front of the fireplace. . . . He crawled beneath it and slept until morning."

The main part of the house, as my parents found it, was about 24 by 30 feet. It was built of hewn logs well fitted together and chinked with lime-clay mortar. Ends of the logs were cut off flush at corners. The exterior was covered with clapboard siding. Wide boards nailed on vertically and covered with wallpaper constituted the interior wall finish. The doors, all handmade, and the trim about doors and windows were of walnut. A deep veranda extended the full length along the south side.

At the east end of the single big room was a huge fireplace that, before my time, had been completely boarded up. Only the mantel and decorative trim, all of wide walnut boards, remained visible. (Old-time fireplaces, wide and deep, with big chimneys, were prodigal wasters of fuel. As wood became scarcer, their use was generally discontinued, and they were closed up to prevent loss of heat from the stoves.)

There had been a time when this fireplace was a most important appurtenance, the center of life and activity in the house. A son of the original owner, quoted in a preceding paragraph, reported that though they had homemade tallow candles and lamps, made by placing wicks in saucers containing lard, members of the family sometimes read by the light of the fire. Often, a quarter of drying beef or venison hung at the fireside. All cooking for the family was done there; corn was parched and potatoes were roasted in the hot embers. In the fire's genial glow the mother spun yarn, knitted stockings and mittens, and sewed garments for the children. There also the father made for the youngsters boots and shoes, using home-tanned leather.

A tall Seth Thomas clock reposed upon the wide fireplace mantel while we lived in the house. (It was transferred to a special shelf in the new house when we moved.) Winding it just before he retired was a nightly chore that Dad never forgot. This clock, with a graceful heron painted on the lower panel of its glass door, which hid the pendulum and the cord-suspended weights, ticked off the family's time faithfully for many years. Towns generally operated on standard time after it was adopted by the railroads in 1883, but rural communities long refused to recognize anything but local or meridian time, always referred to as "Sun Time."

The big room served as formal living room and parlor. In addition to its simple furnishings it had a bed for the use of guests. There, also, young sufferers from childhood illnesses might be confined. The single room above, reached by a partially enclosed stairway that made one right-angle turn, was the family sleeping room. In this log house we children were born. At that time the wild turkeys, bears, and Indians had all disappeared. Only an acre of woodland remained, cut off by a creek, at a corner of the farm.

Attached to the west end of the log structure was a one-story lean-to, about 16 by 24 feet—the kitchen. All cooking was done there on a large, flat-topped, wood-burning stove. This kitchen served regularly as dining room, informal living room and, during much of the year, as the kids' rumpus room. Just outside its south window grew a huge rosebush. In June every year it became a mass of gorgeous yellow blossoms. Day and night their perfume pervaded the house. Bees busied themselves there throughout the long days. We youngsters spent untold happy hours gladdened by the color and the fragrance of those flowers.

With three small children running in and out, Mother must have had a great deal of trouble with flies, in spite of cotton mosquito netting over doors and windows. In dining rooms where groups of threshers were eating, they

64

were often so numerous that several of the hostess' assistants had to stand about the table and, with clusters of paper ribbons attached to long sticks, shoo them away. Mosquitoes plagued us every summer; usually we had to keep smudge fires going if we wanted to sit outside in the evening. No doubt most of these pests came from the rain-water barrels, then our only source of soft water.

The big room of our log house was a wonderful place to play in winter and in other seasons when it was inclement outside. It was cool in summer, having thick walls and good cross ventilation. In winter it was heated comfortably by a box stove (an exceptionally efficient heater) near the center of the room. A smoke pipe rose vertically, passing through a metal thimble in the ceiling above. Some six feet above the ceiling it made a right-angle turn and, after an eight-foot horizontal run, entered the chimney. This long pipe no doubt radiated fully as much heat as the stove itself did.

Being heated by the smoke pipe and by the floor, which, supported by heavy exposed beams, was the ceiling of the room below, the upstairs room was warm and pleasant when we went to bed, no matter how cold the weather. Upon waking some mornings, we boys found that snow, sifting through roof shingles, had collected in little ridges on bed and floor. By that time, however, the room was comfortably warm.

Dad must have found it anything but comfortably warm when he got up some mornings. At night he always had at hand a big, knotty piece of wood that he called a "night-laster." This he put into the stove just before finally closing the drafts. The tough stick held fire well through the night. He could kindle the embers quickly into a brisk fire by throwing in dry wood and opening drafts.

A few steps from the kitchen was the well with its chain pump, which had a wooden pump stalk extending deep into the water. A few steps beyond the well was the milk house, built of hewn, chinked logs, with an overhanging roof to form a front canopy. Milk houses in those parts were always provided with such a canopy, commonly called a "portico." A huge trumpet vine overspread the entire outer wall and the shake roof at the east side. The shade of the vine, combined with the thick walls and the brick floor laid on earth, kept the interior cool. Inside, along a side wall, was a shallow wooden trough for the storage of milk and other dairy products in crocks. All water pumped from the well for livestock ran through the trough to a depth of about two inches, keeping contents of the crocks surprisingly cool.

In the settlement of his father's estate Dad came into possession of a tract of land that included a wooded area of five or six acres. Dad and another man felled all the trees. All suitable timber was cut and split into fence rails. We boys had a hand with the rails, working with the men, using axes, wedges, and mauls just as young Abe Lincoln must have done. In addition to wedges of forged steel we had homemade ironwood wedges and ironwood mauls similar, no doubt, to those he used. In all probability, however, his rails came altogether from large, straight-grained logs, while ours came from comparatively small second-growth white or black ash, much of it so cross-fibered and tough that our rails were quite rough and splintery.

We blasted the stumps with dynamite, burned the brush and the shattered stumps, then plowed the "newground," as all land newly cleared for farming was called. It is questionable whether any job ever involved more strenuous muscular exertion than that kind of plowing did. The ground was interlaced by tough green roots. Many were ripped out by the plow, but the big ones were so firmly anchored that when the plow hit them, giving the plowman a violent wrench, he could do no more than work it free, then go over or around them—dynamite had not budged them.

Near the center of the clearing was a natural conical basin twenty-five or thirty feet in diameter and about six feet deep. Doubtless it was what is known as a kettle hole, formed in the ice age when a glacier laid down at the spot a huge mass of ice with drift surrounding it. Water remained in the hole much of the time, and the ground around it was nearly always wet. At the first opportunity we ran a tile drain in and extended a branch right into that pit. We then scraped in dirt from the surrounding area until only a wide, shallow depression remained. Water never again collected there to remain long.

After dynamiting the green stumps in the newground, we tackled others on the property, about one hundred all told, some of them pretty large. No one regarded this as particularly dangerous work because the dynamite, containing about forty per cent nitroglycerin, was so stable that of itself it could not easily be set off. In the presence of detonating caps, however, it had to be handled with care and common sense.

With a long auger we started at ground level and bored a hole slightly larger than a stick of dynamite, slanting it downward to the center of the stump—into the central tap root when possible. For a large, sound stump a stick or a stick and a half of explosive was used; half a stick was enough for a small one. The charge was primed with a detonating cap and an attached powder-core fuse, then eased gently to the bottom of the hole with a slender stick. Soft, damp clay was pushed in lightly behind it to a depth of several inches. Earth was next tamped in firmly to fill the hole. The fuse was cut off so that two feet or more extended from the charge. Tools and the supply of explosives were collected in a basket and carried several paces from the stump. Finally the end of the fuse was split and a match was applied to the exposed powder core. Then, carrying the basket, the shooter hurried off to a spot some two hundred feet distant. Soon came a muffled explosion. Pieces of stump, particles of soil, and a cloud of dust and smoke were hurled upward, 150 feet or higher. Green stumps were usually split into segments but rarely blown out completely. Stumps whose roots had decayed somewhat were commonly thrown out whole or in big pieces, leaving gaping craters in the ground.

Swarms of honeybees inhabited cavities in numerous trees in the original forest. Beginning early in the settlement of the region, bee-hunting became almost a profession. To locate a bee tree, a little honey was smeared on a convenient tree, and the hunter settled himself nearby to watch. Soon, a worker bee would appear, gorge itself with the sweet, then take to the air, homing toward the treetop hive. It was the hunter's job to keep his eyes on the bee, even though that might mean sloshing through pools of water, running through brambles, tripping over logs, or bumping into trees. Once he saw the bee disappear through a knothole or other opening in a tree, he had the cache located.

His sweet tooth may have been clamoring for immediate satisfaction, but usually at the time he did nothing more than mark the tree, note carefully its location, and spot a few landmarks. He knew full well that felling that tree without adequate preparations to assure completion of the job of removing the honey before nightfall would be playing directly into the hands of numerous eager competitors; given half a chance, bears, coons, possums, or skunks would have licked up the last drop. Bears were known to gnaw for days at the tough wood of a tree to get at a store of honey. Once the hole was large enough, Bruin would thrust in a paw and scoop out the sweet-filled comb, devouring it, together with all bees that might be clinging to it, altogether unmindful of the forays being made against him by the angry insects.

One of Dad's pioneer ancestors was reputed to have located as many as thirteen bee trees in a single day by following with his eye the homing flights of bees from blossoms. Dad could not have qualified as an expert at this art, but he did locate a bee tree that stood on our farm. We boys accompanied him when he went to cut the tree. We could see the bees buzzing about a small hole in the trunk, some thirty feet above the ground. Dad kindled a fire and piled upon it damp, punky wood to make a smudge of smoke. He felled the tree, and as it struck the ground, the trunk broke open near the hole used by the bees, the break within a foot or two of the smoking fire. The bees poured out in numbers, but the smoke quickly subdued them. Armed with a smoker (a bellows-connected metal box containing a smoky fire), we proceeded to take out the honey, some of which was trickling from broken combs. With paddles and ladles we lifted out the sticky mass and transferred it to pails, altogether six or eight gallons. We sampled the store as we worked, eating comb and honey together. Within half an hour I paid a heavy penalty for overindulgence, suffering stomach cramps of extreme severity. Old bee men told us that I probably got an overdose of some vegetable poison that had been in the nectar, perhaps from wild lobelia blossoms.

Some years later, when all the trees had been cut from the plot in which the bee tree grew, we planted the entire area to catalpa seedlings, setting the little trees at corners of four-foot squares. Catalpas were widely sold and planted at the time, to grow into fence-post timber. We cultivated the trees regularly, year after year, until they became too large to cultivate. By the time they had grown large enough for use as posts, the entire grove had become a thick jungle of raspberry bushes, grape vines, poison ivy, and seedlings, as well as saplings of numerous varieties of forest trees. This made cutting a difficult, unpleasant job. The catalpa wood made excellent posts. The actual cost to us, however, all factors considered, was probably much greater than the cost of good commercial red cedar posts would have been.

Our experience in this case taught that Nature, which abhors a vacuum, has no greater liking for vacant space on land capable of growing trees, vines, and shrubs. It is astonishing how quickly natural agencies will reforest any ground that is not altogether sterile and lifeless, provided that it is not disturbed by the plow or other implement, or by pasturing animals. Wind-borne seeds such as those of the elms, maples, box elders, poplars, and the like, float in and germinate quickly, often in spite of well-established grasses, weeds, or other plants. Squirrels carry in nuts and acorns that they plant unintentionally when they dig shallow holes and cache them in the ground. Birds probably contribute more to the process than any other agency, distributing widely seeds of fruit-producing trees, shrubs, and vines.

70

EIGHT

A New Barn and a New House

A FAIRLY LARGE BARN AND a wagon shed, flanked right and left by attached corn cribs, both built of logs and roofed with oak shakes, came as original equipment with the farm. As a boy of about four, I once climbed into one of the cribs and, with commendable intentions, tossed two or three bushels of corn to half a dozen fat hogs, already well fed, in an adjoining pen. This did no real harm, but it made some extra work for Dad, who had to retrieve the over-generous ration of ears. I was given a deserved reprimand, but it was mild, befitting my tender years and the relative harmlessness of the offense. Just the same, it was effective; never again did I overfeed any stock.

After using these primitive buildings several years, Dad decided to build a new frame barn to replace both. At that time a good barn was very important on all farms. Some farms now, having no livestock, need little or no barn space. On many others, barns are still indispensable—they increase the earning power of the establishment. This explains why one sees occasionally a farmer's family living in a small, dingy house while nearby is a large expensive barn. In most such instances the owners are counting on that barn to help earn money that later on will provide a better home and a better living.

We had only a few trees suitable for lumber, but Dad's brother, Uncle Cal, had a large wooded tract. With a bill of materials made out by a carpenter, the two set to work that winter with axes and saws, felling trees that would yield the needed lumber. They snaked the logs together, and, when the snow was deep, loaded them upon bobsleds. They hauled them to a sawmill, marking all with red kiel for identification and to indicate dimensions in which they were to be sawed. I was permitted to watch this work at times, all very interesting, for both were skilled woodcraftsmen and experts at using horses for getting the big logs from the woods to the mill. The sawed lumber was piled near the building site. A shingle mill was set up nearby, and shingles for the roof were cut out of clear white oak.

Early in the spring the foundation was staked out, and field stones, large and small, were put into holes dug for them. Above the ground the stones were carefully laid up and leveled to form foundation pillars. In this way all of the boulders that had accumulated on the farm, and some from a neighboring farm, were disposed of.

Carpenters cut framing timbers to length, chiseled out rectangular mortises, formed tenons, and bored holes for fastening pins. Pins of stout hickory were split out, shaped, and pointed. Sills were laid in place on the stone pillars and pinned together. On the sills other timbers were placed and pinned together in "bents," each bent being so placed that it would be reached in proper order when the structure was raised.

Then came the "barn raising." Several neighbors were on hand to help. The professional barn-raiser came with his

72

spanking team of horses, ropes, blocks, sledge hammers, and pike poles. With his team he raised the "gin pole" to an upright position. This was a strong wooden pole, tall enough to hoist the topmost part of the frame, with ropes and pulleys fixed at its top. It was secured firmly in place by guy ropes. Ropes running over pulleys at the top of the pole were made fast to the first bent, and the horses, pulling at the end of a rope, brought it to an upright position. While men with ropes and pike poles steadied it, temporary braces were nailed on to hold it plumb. I was about five, too young to understand fully, but I was thrilled by this and succeeding operations, all of which I watched, wide-eyed, from a safe vantage point. The second bent went up in the same manner, but it had to be raised more slowly because crossbeams, braces, and nail ties had to be inserted in their places in frame members as the bent approached the perpendicular. When it stood upright, wooden pins were driven in so that beams, braces, and ties held as a unit the two bents now up.

The remaining two bents were raised and secured in place in the same manner. After this the plates (horizontal eave beams) were hoisted into place, their mortises fitting the tenons of upright posts and braces. One of these plates ran along at the "square" at either side, the full length of the barn. They were fastened to the posts, carpenters walking over them and driving in pins with their adzes.

The final operation was to hoist the two purlin plates. They ran lengthwise of the building and were supported by posts and braces so that they came midway between the eaves and the peak of the roof to be. As one-piece timbers of sufficient length could not be obtained, two beams were spliced and pinned together to form each purlin, as well as each eave beam. When the purlins were in place, the carpenters, walking catlike over the sharply inclined faces, drove in the fastening pins. To me this was an exciting, daredevil feat; but to carpenters, as to structural steel workers, standing and walking over slender supports high above the ground are commonplace, all in the day's work. To do it successfully one must have a good sense of balance, complete freedom from acrophobia, and must have accustomed himself gradually to working in such positions.

My father had an abnormal fear of heights that he was never able to overcome altogether. He told me that in his youth he once climbed a ladder to the second floor of a building under construction. Looking down, he became panic-stricken and at once started to descend, crawling slowly, snakelike, over the ladder, desperately gripping each round. I once talked him into going with me via elevator to the top of the Washington Monument. The experience proved anything but pleasant for him. He remained tense and ill at ease until the elevator returned us to solid ground.

Early frame houses were framed much the same as this barn. Their timbers, however, like those of the first frame barns, were hand-hewn, not sawed. Groups of men using ropes and pike poles, raised them into place, fastening joints with wooden pins.

When the last timber of our barn had been hoisted and pinned in place, the gin pole was taken down and loaded with ropes, blocks, and other equipment for transfer to the next job. The raising was finished, but the carpenters' job was then just well under way.

They had to hoist the rafters and nail them in place, apply the rough vertical siding boards to the outside of the

74

frame, and close in the gables to the peak of the roof. Just below the peak they put in a 4 by 4 inch track, full length of the building, on which was mounted a "car" for carrying hay into the mows. The track was made of oak dressed by the carpenters, and was supported by steel rods secured to crossties nailed to the rafters. Next, roof boards, spaced for five inches "to the weather," were nailed to the rafters, and the oak shingles were applied.

Partitions, mangers, and feed boxes were put in for a milking room equipped with stanchions for eight cows, single stalls for six horses, and, opposite these, two large box stalls. The latter were used for mares with young foals, for young calves, or for ailing animals. Heavy oak floors were provided for all areas to be used as quarters for animals. No water was supplied inside the barn.

The floor level of the 48 by 72 foot barn was about two feet above the ground. Earth fill retained by rough stone walls provided approaches to the wide side-entrance doors that opened to the threshing floor, which was about twenty feet wide and open to the roof, extending crosswise through the building at its center. This was a floor designed to support a heavy threshing machine, as it did a number of times when our grain, in sheaves, had been stored in the mows that occupied the upper story at either side. At haying time wagons loaded with hay were driven onto this floor for unloading. There, also, a baler was set up now and then to compress surplus hay into bales. This machine was powered by two horses, hitched to the long lever of a mechanism known as a "horsepower," walking in a circle outside.

The threshing floor was covered by two plies of wide elm boards, as were the wide bay adjacent to the milking room (used for storing machinery) and the feeding alley between the single stalls and the box stalls opposite them. The mows were floored by a single thickness of elm boards.

As an afterthought a few years after the barn was put into use, a space four feet wide was partitioned off from the machine-storage bay to be used as a corncrib. In case bad weather set in before field husking was finished, we snaked in several shocks of corn on a sled and set them up on the floor of the bay. We could then husk at our leisure, unmindful of howling winds, snow, ice, and low outdoor temperatures, and feed fodder as well as some of the corn directly to livestock without additional handling.

With the exception of the big side doors and the ventilating louvers set into outside walls (all made from commercial white pine), every stick in the new barn came from native timber. The same was true of a chicken house, built about the same time. A new smokehouse, a granary, and a buggy shed, with a big corncrib at either side, all had native timber frames, commercial products being used only for roofs and sidings. About 1910, a second, slightly smaller barn and a wood-stave silo were erected on a site 150 yards from the first barn.

"Paint" made by stirring Venetian red into crude oil (as petroleum was commonly called) was applied to the rough oak siding of the first barn. This was intended to do little more than color the boards during the weathering process. It served that purpose fairly well. Years later, because wide cracks opened between the boards, all were ripped off and matched commercial siding was put in their place. This was coated with factory-made red paint.

The oak shingles proved unsatisfactory. Probably, they would have done much better if more time had been allowed for drying and seasoning before they were put on. The wider ones curled badly after a few years; then winds

76

whipped them off, and leaks developed. At length, after repeated patching, Dad bought standing-seam galvanized steel roofing and rented a set of tools for applying it. A brash fellow of about eighteen, I set about putting the new roof over the shingles, working altogether alone. I got the job finished without accident, then installed rain gutters and downspouts, previously omitted.

The overflow of machinery from the barn bay, including generally the binder, mower, hay rake, plows, cultivators, harrows, etc., was stored in a building separated from the barn by a wagon driveway. In this building we also had pens for sheltering and feeding hogs, with a storage loft for corn above them. Our straw was always built into a stack near the barn, at its leeward side. Usually, we had a pole structure in the stack, straw being piled over three of its sides and its top. This enclosure provided winter protection for livestock—as a rule, cattle only.

Not long after we got the new barn, Dad and Uncle Cal cut logs and got out lumber for the frame of our new house. Most of the men who worked on the building ate with us at the family table, as was generally customary at the time (the barn workmen did too). Some even had to be provided with beds. The plasterer, with his Negro helper, ate and slept at our house until all of their work was finished. They started early and worked late every day. Once, about an hour before dawn, the boss aroused his assistant and bade him hurry with his dressing. The young man shortly appeared, yawning and rubbing his eyes. Noting the darkness outside and the steaming breakfast on the table, he exclaimed: "Man! Two suppahs in one night!"

Unlike our sister of preschool age, we boys never knew just how we got moved from the old house to the new.

One morning we ate breakfast and dressed in the familiar home surroundings, then went to school. When we returned, we found the furniture arranged in a strange, alien sort of house, new throughout and smelling of new wood, paint, and varnish. This house, with eight big rooms, provided a generous amount of living space. There were improvements and advantages that we did not have before. Likewise, as we discovered with the passage of a little time, there were some disadvantages.

As winter came on, we noted that it was not as warm as the log house. As in all frame houses built at the time, no sheathing was applied to the outer walls; the weatherboarding, of dressed, matched lumber, was nailed directly to the studs. No one had yet thought of using building paper or other insulation in houses as barriers against the transmission of heated air. In the lower rooms floor boards were nailed directly to the joists, without subfloors or flooring felt. Since there was a large volume of unheated air under them, the floors were always cold in winter; the stoves we had could not keep them warm.

There was a big volume of attic space above the upstairs rooms, but this had no noticeable effect in preventing the loss of heat in winter. In summer the attic air absorbed the sun's heat, which radiated freely to the rooms below, not only in daylight hours but during much of the night. My brother and I stoutly maintained that our upstairs room was the hottest in the house in summer and the coldest in winter. Probably it really wasn't, but no one seemed to think it worth while to argue the matter with us. Some breezeless nights in midsummer were so hot that, after sweating wakefully in bed a while, we had to give up and stretch out on the floor of the room below that, shut off from attic heat, was comfortably cool.

78

Going to bed in winter seemed to us at times like an expedition to the Arctic—there was no heat in the stairway, the long hall, or in any upstairs room. A thick layer of blankets and comforters was provided for our bed, but it happened rather often that one or the other of us would unconsciously roll himself into the lion's share of the covers, leaving his partner chilled and wakeful. This naturally was conducive to some rather tart exchanges. Getting up in that frigid room was an ordeal that we would have always postponed indefinitely; but it was a fairly strict family rule that all must be on hand when breakfast was ready, usually not later than six A.M. throughout the year.

The roof of the new house, of rather steep pitch, was covered with slate shingles. Always, after five or six inches of snow had piled up, it would avalanche to the ground. The accompanying roar was quite startling at first, but we gradually became accustomed to it and finally came to pay it little attention. It is a wonder that no one was ever buried under a big slide.

This fault of the roof was fully offset by qualities that kept water that fell upon it quite clean. All roof water was stored in a cistern with a pump, near the house. It was a great satisfaction to have an ample supply of clear, soft water—no more rain barrels with their attendant mosquitoes. A good many years passed before anyone thought to pipe cistern water to a pump inside the kitchen, with convenient sink and drain.

Like most of our neighbors, we now had a big parlor. Here were our best carpets, our best furniture, and a bookcase-desk in which reposed our best books. In a prominent spot was our parlor organ. The reed organ was then commonplace in farm homes, but there had been a time, about a generation earlier, when it was something of a status symbol. A story was told of an unwashed old fellow in the region who, after buying one, remarked to neighbors that in those times "everybody who is anybody just about has to own an organ." This gave rise to no little merriment, for the man was widely regarded as an inept, lazy incompetent, rather low on the totem pole.

On a center table in our parlor was a large family Bible with space for all vital family records and with the names of my father and mother printed in gold letters on its embossed leather cover. A similar Bible was to be found in almost every other home in the community. Near the Bible lay several photograph albums and a stereoscope with an assortment of stereo photos. There stood also our biggest, most expensive lamp, not unlike lamps in numerous other homes. All of these ornate lamps had tubular wicks to increase the air supply to burners. Since all tended to be cumbersome and more ornamental than useful, they were generally used only on festive occasions.

For years kerosene lamps provided the only illumination in farm homes. Candles were passé—we never saw them used except on Christmas trees. Cleaning chimneys and burners of lamps and lanterns was an unpleasant daily chore. Not even the best oil lamps produced light that was altogether satisfactory for reading or close work. In time, wick-fed kerosene lamps with incandescent mantles appeared; they gave good light. Some preferred them to the acetylene and gasoline lamps that were gradually coming into use, considering them handier and safer.

Acetylene lamps, for some time the best available for automobile lighting, were used only to a limited extent in homes. Some rural churches had acetylene or gasoline lighting plants, generally with complicated systems of pumps, generators, valves, and piping. When they worked, they gave excellent light; but they were not altogether dependable or predictable.

A few families, including one near us, installed small electric systems specially designed for farm lighting, with low-voltage generators driven by gasoline engines and generally with batteries for the storage of current. They served well when in good operating condition, but at times the lights would flicker, then wane to a mere red glow. Occasionally, they failed altogether just when they were needed most.

At times men and boys might have voted lanterns fully as important as lamps. They were indispensable in the short days of winter when chores about the barn, consuming about an hour each morning and each evening, had to be done. Every family had one or more. A lantern was always carried when one went anywhere in the neighborhood at night, to avoid mud puddles, snowdrifts, or obstructions.

For several years our parlor was used only when we had parties or special guests. Gradually it came into more general use, and eventually it was thrown open to become essentially an annex of the living room. This change was encouraged by the installation of an improved stove in the place of the woodburning "Round Oak" heater. The new stove was a "base-burner," so called because the anthracite coal fed down from a hopper at the top and burned at the base of the firepot. It gave a cheerful aspect to the room day and night, because glowing coals and the lambent flames were visible through mica-covered openings in doors on three sides. It radiated a great deal of heat, but it had little effect in warming floors.

We also set up in the kitchen a new range made to burn either wood or coal. Its reservoir for warm rain water was larger than that of the old stove from the old kitchen. Its oven threw out a great volume of heat, a perfect place for warming cold feet and a powerful temptation to put off doing outdoor chores in winter. Near this oven plants for the garden were started in early spring. There also, occasionally, frail baby pigs, lambs, or chicks, snugly wrapped in blankets, basked in comfortable warmth until they became strong enough to cope with conditions in the outside world.

A few steps from the kitchen door we built an upground cellar, its inside arrangement similar to that of the old milk house; also, a building that served as combination summer kitchen, woodshed, and laundry. Every spring Mother moved cooking utensils into this building and did all cooking there until cold weather came. Almost every farm had a summer kitchen at the time. The object was to keep the main kitchen, which usually doubled as dining room, cool and free from cooking odors. The arrangement was also supposed to reduce the fly nuisance.

Few country people had known much about gasoline until, almost overnight, gasoline kitchen ranges came on the market and quickly caught the public fancy. A man in the neighborhood made a swap and got one of the new stoves for his wife, one of the first to appear. "Sure is funny," he reported, "to see her a-standin' there, a-cookin' without no fire."

Mother bought one and had it set up in her summer kitchen, to be used there only. No one ever quite overcame the feeling that the thing was a little dangerous, the volatile fuel being stored in a tank near the burners, to which it flowed by gravity. A decade or so later kerosene-burning ranges crowded the gasoline stoves out completely. The transition, which came within a short time, apparently was induced by the belief that kerosene was the safer fuel; at the time the price differential was too small to be important.

A well drilled at a point convenient to both the summer kitchen and the house tapped a good supply of water at a depth of sixty-five feet, forty-five feet in the bedrock. Surface water was "cased off" by a pipe driven down thirty feet. The water was cold and clear but quite hard, with a pronounced sulphurous taste and odor. Some in the community would have called in a "water witch" with a peach-twig divining rod to tell them where to drill, but dad took no stock in divination. He and Mother simply decided that a well would best meet requirements at a certain spot, and there it was sunk. No doubt drilling to the same depth anywhere else on the farm would have been just as successful. Furthermore, regardless of its location, the well in all probability would have yielded sulphur water, for water from most drilled wells thereabouts contained some sulphur compounds.

Until about the time this well was drilled, the water supply of most farms came from "dug" or open wells, walled with brick and provided with wooden covers. In most cases the water was clear, cold, and tasteless. No doubt most of it was pure and safe for drinking, since it flowed in from sources deep underground. But such wells were readily subject to contamination from surface sources. Doubtless, contaminated well water was responsible for many of the cases of typhoid that occurred.

Occasionally at threshing time as much as 1,000 gallons of water were pumped from our old open well within a few minutes. This never lowered the water level perceptibly. After years of hand pumping, we erected a steel windmill over this well. It saved a great deal of time and labor because even a light breeze would keep the pump going, supplying water to slake the thirst of our stock, really prodigious during much of the year. At that time wooden windmills for pumping water and, in a few instances, for grinding feed, dotted the landscape; but all new mills that appeared were made of galvanized steel.

A neighbor had an open "gum" well in his woods to supply water for his cattle. The gum was a hollow sycamore log with an inside diameter of three feet, set twelve feet into the ground. The well was fed by water that seeped in from the surrounding undrained soil. Hollow sycamore logs were commonly called gums by local people, whether used for casing wells or set up and roofed over to serve as smokehouses or as shelters for pigs or chickens.

A story was told of a young husband and wife who had both been killed by "damp" encountered in one of the early wells dug in the area. When the man failed to appear at meal time, members of the family went to the well site and found him lying at the bottom of the excavation, about fifteen feet deep. Fighting off those who sought to restrain her, the wife ran down the ladder and at once fell over unconscious. Both were dead when rescuers, working frantically, finally got them to the surface.

Damp in wells was usually marsh gas, or methane, produced by the decomposition of vegetable matter in the glacial drift. Well-diggers were reported to have found tree trunks and limbs buried deep in soil never before disturbed by man. Such deposits were referred to as "Noah's brush-heaps" or "Noah's barnyards." They were found in what geologists call the older drift—the deeper glacial deposits.

To be sure of having safe water, most farmers in later years had wells drilled into the bedrock. A galvanized steel pipe about four inches in diameter was driven in, extending from the surface to a depth of ten feet or more into the underlying solid rock, to seal out all water except that coming from veins or crevices in the deeper rock strata. Contamination by surface water is therefore impossible so long as the casing pipe remains sound. Because the water in these wells is usually strongly impregnated with sulphur compounds, highly corrosive to steel, there is some question as to how long an effective seal will be maintained.

After the family was ensconced in the new dwelling, the old log house became a workshop, with carpenter's bench and an assortment of tools. It was used for storing everything that didn't seem to fit anywhere else and for all indoor work connected with butchering. In a corner of the big main room, I built for my own use a small photographic darkroom. The old box stove remained in place, providing heat when needed.

Built as it was, termites would have found it ideal for all their stealthy activities. Fortunately, up to that time none had appeared on our farm or elsewhere near. With reasonable attention and care it could have been preserved indefinitely in sound condition. As time went on and the need for it diminished, however, maintenance was neglected. The original supporting pillars of oak gradually rotted away; then the lower logs began to decay. This caused uneven settling, throwing doors and windows out of plumb and allowing lower floors to sag and rot. At last, a useless eyesore, it was torn down. The logs of its walls, which had so long afforded protection against the elements and given friendly seclusion from the outside world, met the fate of logs from the old barn, being cut into firewood.

NINE

Rescued from Seas of Mud

OUR HOUSE STOOD ABOUT one-eighth of a mile off the highway. Buildings on about a third of the other farms in the area were also set well back, in some cases half a mile. Some chose this arrangement because the remote spot afforded a better building site than one nearer the road; others probably were motivated by a desire for seclusion. In a few instances, it appears, the building site was chosen before there was a road.

One advantage of the deep setback, we found, was that road dust, always abundant in summer, bothered us very little. Another was that a fairly large proportion of tramps and peddlers, deterred by the extra walking, passed us by. A lane, which before we piked it might become impassable at times in spring, led from the road to the house and extended to the barn.

Beyond the picket fence that enclosed our lawn, a wide open plot extended to the road. One day in early spring, Mike Geary, a dealer in threshing outfits, ran into some difficulty with a big steam engine that he was driving on the highway. Dad gave him permission to leave it temporarily in our lot. The ground was so wet and soft that the drive wheels of the heavy machine cut long furrows a foot wide and nearly a foot deep. "Bill, I'm sorry about that," said Mike, as the two men gazed ruefully at the ugly gashes, "but I reckon neither of us will worry about it fifty years from now." No one had real cause for worry at that time or later, although the damage did appear quite serious. Some weeks later, the engine gone, we closed the wide furrows with shovels. Within a month rains and growing grass erased them completely.

Not long after we moved into the new house, the work of macadamizing the roads of Ridge Township got under way. Until that time public roads—even streets in town— were unpaved. All of these thoroughfares served fairly well when dry, which was the greater part of the year. Spring thaws and rains, however, turned them into quagmires, in spite of periodic grading and reasonably good drainage. Travel over them then was practically out of the question except for those who walked or rode horses. One could get through in a two-wheeled cart if he had a good horse and was neither too impatient nor too squeamish about being liberally spattered with extraordinarily sticky mud. Even with the limited traffic, the mud attained a tough, rubber-like consistency. A horse's feet would sink in ten inches or more, and when he lifted them, one would hear a resounding smack as the vacuum was broken.

In pioneer days, before much drainage work had been done, families moving in sometimes got bogged down completely in the big mudholes that dotted primitive roads in the Black Swamp. It is related that some of the settlers living near the worst spots took pains to keep them impassable by ordinary means. They maintained yokes of

90

stalwart oxen for the sole purpose of aiding mired-down travelers—at a price. The owner of a tavern near an exceptionally wide and deep mudhole that he never allowed to dry up is said to have exacted from the man who bought the hostelry from him an extra for the "travelers' relief" concession.

With plows and scrapers, all of course horse-drawn, roadside ditches of our township were widened and deepened by our road builders. This was done to provide better drainage and to get material for higher, drier roadbeds. Bridges and culverts were repaired or replaced as necessary. Then crushed limestone, uniformly about the size of a walnut, was dumped inside plank forms on the graded roadbed. The stone, leveled to a depth of about eight inches, had a width of about twelve feet after the planks were removed. Constructing roadbeds and hauling stone provided for local people a great deal of employment, regarded at the time as lucrative—a man with a good team of horses could earn about $4 in a ten-hour day.

No water was applied to the stone, and nothing was done to compact or bind it. When the job was finished, there were two traffic lanes, the pike with sloping sides and, beside it at a lower level, what everybody called a "dirt" road. The stone was so coarse and loose that, except in passing someone else, no one ever drove on it unless the dirt road was quite wet and muddy. For that reason it took several years for regular traffic, aided by snows and rains, to get the stone ground down and firmly settled. Even then, the dirt road was preferred when it could be used because it was quieter under wheels and much easier on horses' feet.

A picket fence extended the full length of our farm along the road. With its red, wooden slats, it was identical with the fences now erected each fall to serve as snow barriers along highways at points subject to heavy drifting. Often after a big fall of snow, westerly winds built up deep drifts just over the fence, inside the road. From time to time more snow was added to the drift so that an accumulation, the full four-foot height of the fence, might be piled up from one end of the farm to the other. Deep drifts also formed beside rail fences. Freezing after a light rain or after a slight thaw produced surface crusts on which we could, and often did, walk. We also found conditions then ideal for tunneling and exacavating in the banks.

One time, after an exceptionally heavy fall of snow, the wind drove it well beyond the line of customary lodgment next to the fence and piled it three feet deep over the pike. Early the following morning, a dozen men of the neighborhood, armed with scoop shovels, set to work and opened a lane wide enough for traffic to get through. At the time it appeared that they had assembled spontaneously because each recognized that the work had to be done—telephones for calling them together were not available. It could have been that all or part of them had been somehow summoned by the township trustees; at the time each adult male was required to donate one day of road labor per year.

The establishment of a rural mail delivery route came soon after our roads were piked. Our address became R.F.D. No. 1 instead of Box 224 at the post office in town. Next, farmers began to consider the advantages that telephone service would give them. After several years of

92

dickering, the telephone company extended lines from Van Wert, and we were put on a party line that served about a dozen farm homes.

We had a wall-mounted instrument with a crank that we had to turn to call a number. Visiting and gossiping were indulged in from early morning until late at night. Eavesdropping became a popular neighborhood diversion. If one wanted to make a call on an outside line, he had to lift the receiver, wait until those using the line had finished their conversation, then immediately crank vigorously to gain priority for a call to the operator. As time went on, the service was improved and modernized. Present subscribers are on party lines; but the number served per line is much smaller, and the instruments provided are of late design.

Improved roads, which facilitated the hauling of equipment and supplies, stimulated prospecting for oil and gas in our area, which, some believed, might prove to be a part of the rich Lima field to the east and south of us—for several preceding years a beehive of activity. Within a short time speculators and would-be prospectors had nearly all of the land about us under lease for oil and gas operations. Leases ordinarily ran through a term of several months and provided for necessary easements. The landowner was paid an initial sum, called a "bonus," and was guaranteed a share of all revenues that might come from wells on his property, generally one-sixth. The bonus might be a very tidy sum in the case of a farm near a productive well. "Gushers" that flowed a long time were put down within six miles eastward of our farm. Southward about ten miles there were even more, most of them big producers.

Farmer owners of land on which bonanza wells were located became wealthy almost overnight.

Oil wells in the region were drilled to a depth of about fifteen hundred feet. The tall derricks, all of rough lumber, were .put up by "rig builders" at well sites. Power was supplied by steam engines with boilers fired by coal or by oil or gas from a nearby well. Gas that came from oil wells was often treated as a waste product, being piped some distance from the well to feed an open flame that burned day and night. Occasionally farmers piped gas into their homes for lighting and heating.

A well was sunk about a mile east of our farm and another about the same distance west. Some oil and a little gas were found in each, but there was not enough in either case to be profitable; so plans for further prospecting in the vicinity were given up and leases were surrendered.

Some of the wells that started as big producers continued to yield oil in paying quantities for years. At last, however, almost all in the region failed and were abandoned. The derricks, the tanks, the pumping jacks, the central power plants with their gas engines, the pipe lines, and the big oily blobs over the ground all disappeared, and the lands were restored to their former function of producing crops.

A man in one of the big fields told me that abandoned wells were sealed by pushing a wooden plug deep into the bore, pouring in a quantity of rock cuttings, and finally dropping in an iron ball. Salt water is not known to have escaped from old oil wells to ruin nearby water wells, but some farmers suspect that the deep wells have played a part in the failure of their water wells by draining the veins that originally fed them.

A pipe line carrying gas from fields south of us ran along the road past our farm. Soon after we moved into the new house, a tap was made into this main, and gas was piped in to the kitchen stove. There was a regulator, but the pressure fluctuated from very low to moderately high. The fuel was supplied at a flat monthly rate that would be considered only nominal now. The pressure at last became so low that we disconnected the pipe and went back to the use of wood.

TEN

Animal Companions and Aids

THE MAJORITY OF FARMERS
in the area showed a preference for general-purpose horses,
big and strong enough for effective work, yet not too
heavy or clumsy for other uses. A few were proud owners
of heavy draft animals to which they gave painstaking
care. Occasionally, an individual or a well-matched team
was specially groomed for exhibition at the county fair.
The heavy breeds most favored were Percherons, Belgians,
and Clydesdales. One saw rather often light, fleet-footed
steeds used for carriage work or for the saddle. The best
of them were classified as Morgans, Thoroughbreds, or
Hambletonians, probably few, if any, being purebred. A
number of farmers kept breeding stock and counted on
the sale of colts for a part of their yearly incomes. Few
showed much interest in pedigrees; but no doubt there
were some horses, principally in the heavy draft class,
that were eligible for registration. Most of the horses in
regular use were mongrels, carrying blood strains of several
different breeds. Practically all farm ponies were half-
breeds. No farmer we knew kept mules.

For many years nearly all sons of farmers regularly
became farmers; the path to other occupations was rather

difficult and restricted. Naturally, a few, lacking natural aptitudes, found much of the work they had to do boresome, and they fell short of real success. Often, one could spot these fellows by noting the way they got along with horses. If they feared the animals and failed to understand them, they had to worry along with awkward, ill-trained nags that were hard to handle and caused them much vexatious trouble. Consequently, a considerable part of the work they had to do was, for them, highly bugbearish. One farmer we knew was so lacking in ability with horses that he was seriously handicapped. They worried him, and he was afraid of them. He came along one day, an apprehensive look on his face, driving a clumsy young horse that he was "breaking." When he stopped, the stolid creature stood listlessly, as if half-asleep. Someone remarked that the colt appeared tame and quiet. "He's not foolin' me," replied the uneasy owner. "Don't trust him at all. Right now, I'm sure, he's a-meditatin' some low-down deviltry."

Horses have mannerisms, tricks, and habits that give to each a distinct individuality. Some handle their feet gracefully and briskly; others move clumsily and lazily. A pronounced lazy streak is by no means unusual. Some resort to clever tricks to gain time for resting or loafing or to make things easier for themselves. Bonnie, a little mare we owned, had an ideal disposition and was very dependable; but, hitched to a buggy on the road, she was much given to goldbricking; instead of trotting along at a good clip, of which she was easily capable, she chose, if the driver permitted, to drop into a jogging dogtrot, a pace that could be maintained with a minimum of exertion.

No matter what work might be under way, a sturdy team that we used a long time would walk back and forth across the field at their own mutually satisfactory plodding gait. But at the first sound of the dinner bell both would prick up their ears and, with one accord, they would start "on the double." In spite of anything the driver might do, they would maintain that pace until they reached the side of the field nearest the barn. Unhitched, they would head straight and fast for a hearty drink and their midday meal. After dinner they would settle into their easy-going pace and hold it until the bell pealed the hour for supper. All of our horses learned that the bell meant a break in their labors, but none responded quite so eagerly as this pair.

We had horses that displayed a real sense of humor. Two or three, for instance, delighted in taking a playful nip at an ear or finger of a person who happened to stand unwarily near their mangers. One of these equine jokers would occasionally pretend to bite the seat of one's pants as he bent over while currying him or adjusting the harness. These were all gentle, harmless gestures, plainly intended in fun.

I once bridled a gentle, fleet-footed old mare and led her to the pasture gate, meaning to ride. She stood quietly while I climbed the bars; but the instant I sprang for her back, she leaped forward and I, greatly surprised and chagrined, landed on the ground. Her expression when I approached her after that spill was unquestionably one of amusement at having tricked an upstart young rider.

A little half-pony we had seemed to delight in playing repeatedly a little trick all her own. One might be riding her at a brisk gallop when suddenly, without warning or any discernible reason, she would leap to one side. The rider's inertia of course carried him on in the direction he had been going, and he suffered a rude, unplanned-for landing.

One summer night I was driving a young lady home when a violent storm arose. The wind blew a terrific gale. Rain poured down in torrents, and menacing stabs of lightning streaked across the sky. It became so dark that my horse Jack, the road—all surroundings—were blotted from sight. I gave Jack his head, feeling sure that he could keep to the road better without my help than with it. Far less scared than I, he walked ahead confidently, keeping the buggy squarely on the road, safely away from the deep, water-filled ditches on either side. We had covered a mile or more when Jack stopped suddenly and stood motionless. I clambered out, groped forward in the inky darkness and found that a big tree, blown over by the wind, lay across the pavement. When the storm subsided, I discovered that the tree lay diagonally so that it did not fully block the way. Leading Jack, we squeezed safely through the narrow space between the top of the tree and the ditch and proceeded on our way.

Most horses like to be curried and brushed. Wise horse-men regard this as essential, but many farmers tended to neglect it, especially in cold weather. In spring their horses resembled crustaceans, a thick, hard shell of manure clinging tenaciously to the hair over large areas of their anatomies.

One of our horses somehow got the name Bum, which he by no means deserved. We never named a horse Joe because we knew that horses mistake the sound for "whoa" —at least they pretend to—and stop when no stop is wanted. Dad one day brought home a husky young gray that he had bought. We boys at once named him Ephriam—Eph for short—after an odd character who a short time before had helped us in the harvest field. Eph's I.Q. was not the highest, but he became a valuable worker on the farm.

100

His back was broad and well cushioned, but he had a habit of setting down his feet so solidly that anyone riding him got a hard jolt at each step. Once, two cousins from town were vacationing at the farm. As both boys were fond of horses and eager to ride, the four of us decided one day to go riding. We mounted and set out, each on a horse of his own choosing, all riding bareback. After our steeds had covered three or four miles at a slow walk, the boy on Eph began to think that maybe he had been wrong in regarding horseback-riding as such a pleasant diversion. By the time we got back to the farm, after a jaunt of about ten miles, his hankering for riding had faded out completely. For several days he was more comfortable standing than sitting.

In summer we watered and fed our horses at the end of their work day, then turned them into the pasture. It was fun to watch them as, lying on the ground, head and legs extended, they stretched and rolled in great evident enjoyment. Some easily flipped themselves over repeatedly from side to side; others were much less successful. Naïvely, we youngsters long believed an old saying that each turn from one side to the other registered $100 in the animal's value. When we went to the pasture in the morning to get the horses, we always carried with their halters a bit of corn for each. This paid important dividends because at the first whistle all unfailingly came to the gate, and each, munching his corn, would co-operatively stick his head into the halter.

Dad taught us that in very cold weather the steel bit of a horse's bridle should either be warmed or dipped in water to coat it well with ice before it was inserted in the animal's mouth. Otherwise, mouth membranes would freeze to the cold bit and be torn off. I easily understood the good sense and the humaneness of this procedure because, in an ill-advised moment when I was quite small, I had applied my tongue to a steel door hook in zero weather with extremely unpleasant results.

Because he liked and understood horses, Dad generally got along well with them. He had effective treatments for colic and other common horse ailments. But if a splint, a ringbone, a curb, or a spavin developed, he was helpless. For that matter, professional veterinarians were too. He was quick to recognize desirable qualities and could spot most blemishes at a glance. Windsuckers and heavers rarely got by him. He was not easily fooled as to an animal's age, even by a gyp trader's "doctoring" of the teeth. I suspect that much that he knew about horses had been learned the hard way; a large number of professional traders and dealers were rascals and crooks of the first water, and some of the tricks they resorted to for cheating the unwary were diabolically clever.

In spite of his knowledge of horseflesh, I recall two times when he was taken in. The first was when a professional buyer and trader came along with a sleek little nag that he offered for sale or trade. At the time Dad had a horse that he disliked, so, after some parleying, they swapped. Within a few hours we discovered that the new mare had the heaves, a horse disease in which the air vesicles of the lungs are permanently distended, causing a persistent cough and a characteristic heaving of the flanks. The animal had been doped enough to cause a temporary sub-sidence of all symptoms. This deal, embarrassing enough to Dad as it stood, was made even more so a day or two later. My brother, driving another horse hitched to the family carriage, ran into the dealer. "Why," said the brazen

102

fellow, "not driving your new horse? How come?" "Aw," replied the guileless lad, "she's got the heaves."

Another time a band of gypsies stopped and offered for trade a handsome young sorrel built for speed and free from blemishes. Dad led out a horse he wanted to unload, and a trade was made. Our suspicions were aroused that night when we led the new horse to water. He made a sudden lunge and dashed through the door like a bullet from a gun. The following morning, my brother, warily currying the horse, stopped suddenly, stared a moment, then called out: "Dad, come and look at the readin' on this horse." Sure enough, high on his right hind leg was a brand that identified him unmistakably as a "Western." Somehow the nomad traders had hidden the characters burned into his skin and had done the job so skilfully that they were not noticeable at a casual glance. (To horse-men and farmers of the region all horses imported from ranges of the West were "Westerns." Almost invariably they were so wild, so nervous, so intractable, and so utterly unpredictable that it was impossible to tame or train them to a point where they could be used with any degree of satisfaction or safety. Doubtless a good rider of rodeo caliber could eventually have converted them into pretty fair saddle horses, but no one thereabouts was willing to risk his neck in such an enterprise.) Gingerly we hitched our Western to the wagon beside a steady, trustworthy horse. He leaped and reared time after time; he kicked, plunged forward, then suddenly fell back. Without the restraining influence of his well-trained mate he doubtless would have staged a devastating runaway, kicked someone to death, or broken his own neck. At length he balked and stood as stubbornly in his tracks as if he had been rooted

103

to the ground. We coaxed and wheedled, we urged him gently, and we used force. Aided by the other horse, we got him going again, by crazy leaps and lunges. Then he stopped in another dead balk. We tried everything in the book but got nowhere. Realizing that even if it were actually possible to break and train him to work the process would be too costly in time and effort, we returned him to his stall, and the next dealer who called took him away.

One evening all of us were riding homeward in the family carriage behind Dick, a big sorrel that Dad had just bought. At the sight of someone riding a bicycle, the horse, in wild panic, suddenly leaped sideways and, in spite of Dad's skillful maneuvering, almost dumped us all into a roadside ditch. Thereafter, we scrupulously avoided using Dick as a carriage horse. Many horses at that time panicked at the sight of a bicycle. The machines were used in considerable numbers on the roads, but most riders courteously stopped when horses showed fright and aided drivers in controlling them.

An odd wagon-like contraption "whizzed" by our farm one day at about two miles per hour. Seated near the forward end, the driver propelled it by turning a crank vigorously with both hands. This vagrant fellow, traveling with bed and cooking equipment in his canvas-covered vehicle, was the first of the tourist tribe in those parts. His machine, the first horseless carriage on the local scene, could have been considered the forerunner of the engine-powered automobile that was to come a few years later and frighten most horses out of what wits they had.

The first automobile to appear in our community was brought in as a main accessory of a traveling medicine show. It was so big and so ugly that no one could justly

104

blame a horse for fearing it. Soon, other machines came, in motley array and in ever-increasing numbers. Our horses, like horses everywhere else, seemed to regard them as malevolent monsters with the most sinister designs against them. At the sight of one, even the most sensible and trusted of steeds became instantly obsessed by a mad desire to get away quickly, in any direction, by any means. It was practically impossible to control them. This made the use of horses on the roads exciting and, in many instances, hazardous.

Most of the cars in the early days were driven by doctors and local businessmen. Practically all showed utmost consideration for those who went on the highways with horses. A relatively small number of smart-alecky, cocky auto drivers made most of the trouble for other users of public roads and streets. They seemed to take an insane delight in scaring horses, splashing mud on pedestrians and forcing them to leap or run to escape from the juggernauts. Due to the behavior of these fellows, public hostility against the automobile was aroused everywhere. In this connection may be mentioned the "invention" of a local genius. His ire had been aroused to such a pitch that he considered attaching an extra-long scythe blade to the rear of his buggy. This was to be so arranged that it could be quickly swung to a horizontal position at either side of the vehicle for the purpose of "mowing down" discourteous road hogs when he encountered them.

A few horses had a mortal fear of threshing engines and locomotives. The sight of an unfamiliar animal, especially a large one, was terrifying to many. According to a family story, my grandfather, homeward bound one day with a heavy load of drain tiles, met in the road an itinerant showman with two bears on leashes. His horses immediately bolted in wild panic. Unmindful of his sawing on their bits, they ran at top speed, the wagon bouncing and clattering behind them. Finally, after their mad flight had covered several miles, he managed to steer them, head on, into a rugged fence. That stopped them, panting and trembling, one on either side of the fence. He had to act quickly, slashing harness straps with his knife, to save them from strangulation. Man and horses were unscathed; it is questionable whether the tiles fared as well.

There was always a market for horses; one could sell or buy one whenever he wished. Few kept an unsatisfactory animal long. We always had an affectionate regard for a sensible, faithful horse and never saw one quit the farm without a twinge of sorrow and a fervent hope that he would fall into the hands of a kind, considerate person.

The principal objective in keeping cattle was the production of good beef animals. Average herds numbered perhaps ten or twelve head. Favored breeds were the Shorthorn and the Red Polled. Often there was a Jersey or two—in some cases Guernseys—in the herd for the purpose of insuring a richer milk output and butter of a higher quality. For a long time butter was the only dairy product that had any market value.

Sheep were kept somewhat sporadically—rarely did one see them regularly, year after year, on any farm. Flocks generally were kept rather small, mainly, it appeared, because of marauding dogs. We had sheep on the farm only about half of the time, ordinarily not more than fifteen ewes. Breeds most common were Shropshires and Merinos. Shearing was done with hand shears (usually by a man who specialized in that work), and the fleeces, unwashed

106

and tied with a special twine, were sold soon after. Few lambs were marketed until they had grown almost to full maturity.

Considerable caution was exercised in growing hogs because of the danger of hog cholera. Before the development of an anti-cholera serum in 1905, the disease occasionally wiped out entire herds within a short time. Generally, we had about three brood sows. We kept our hogs healthy by sanitary measures that included frequent changes of feeding pens and grounds. Breeds most often seen were Berkshires, Poland Chinas, and Chester Whites. Most farmers gave careful attention to the selection of all breeding stock, but there were few purebred animals.

Nearly all livestock was sold to professional buyers, most of whom were expert judges of quality, and could estimate weights with surprising accuracy. They assembled the animals in pens adjacent to railroad sidings and shipped them out to city markets.

Several of the horses and a few of the cows with amiable dispositions, exceptional intelligence, and tricks that we thought cute became special friends and pets of the family. Among them was a Jersey cow, so friendly and gentle that she was long a favorite. To the delight of visiting children, she allowed them to ride comfortably upon her back by the hour. A few bottle-fed orphan lambs became pets—and eventually nuisances. Like the lamb in the nursery rhyme, they delighted in following us wherever we went, though never to school.

My brother once bought from a neighbor a scrawny runt pig, which he installed in a pen specially built for him. Recalling the story of the handwriting on the wall, as recorded in the Book of Daniel; he named the pig Tekel because, he said, if weighed in the balances, he would be found wanting. Tekel got the best possible care and an abundance of the choicest pig foods. He slept, snug and warm, in a little house provided exclusively for his personal use. Soon, a curl formed in his tail to become a permanent feature, and he began to grow thriftily. He became a conversation piece, an object of prime interest to visitors, nearly all of whom called him "Tekiel." Tekel endeared himself to his owner and to all in the family. So much so, indeed, that by the time he had grown to mature hoghood and it became necessary to sell him to the butcher in town, we all felt sorrow and remorse, as if, in a way, we had become accessories to the murder of a friend.

We never went in for caged pets. I did one time make a wire-barred wooden cage with a large glass window in which I imprisoned a mouse taken uninjured from a trap, meaning to study it and learn something of its habits. A little later I caught a brownish, white-bellied, long-tailed wood mouse. I put him into the cage with the first specimen, one of the species commonly found in barns and houses. This, I soon learned, was a tragic mistake; the larger, more aggressive wood mouse attacked and slew his fellow prisoner. A day or two later he gnawed a hole in the wooden wall and made off for parts unknown.

We had no goats, but we had abundant opportunity to observe and study them. A man who lived near the school kept half a dozen, and we saw them twice daily as we went back and forth. At times we found one or two, like circus performers, walking nonchalantly along the top rails of the fence. All liked to climb to the top of the strawstack and stand there, like Barbary sheep on a mountain peak,

108

placidly gazing over the landscape. Often, an old billy would climb to the roof of the low barn. He seemed to derive a world of satisfaction from standing at the very peak, his long white whiskers waving rhythmically up and down as he chewed his cud and meditated.

Cats serve a useful purpose on the farm, preying night and day upon mice and rats; but for us it was always a case of too much of a good thing. The natural fecundity of these animals alone would have assured a cat population several times as large as we actually needed, but the number was artificially augmented with a fair degree of regularity by the foisting upon us of surplus cats from other farms. It was a common practice for a family, finding its current stock of cats too large, to collect a gunny sack full, load the bagged tabbies into the family carriage, and turn them loose miles from home. It was nothing to see a new cat or two appear among our regulars almost every week and settle down contentedly, adopting us without reservations, particularly after they learned that they could count on being well fed every day. Although direct evidence was lacking, we were certain that all those cats had arrived via gunny sack. From time to time we found ourselves with so many cats that they were a bothersome nuisance. Then we borrowed a leaf from the neighbors' book, weeded out the least desirable ones, bagged them up, and carted them away to find new homes.

We were all too chicken-hearted to kill any except in the case of one now and then that appeared incurably sick and therefore better dead than alive. Some of the boys we knew had no qualms whatever about using direct action for ridding themselves of unwanted cats. They thought nothing of shooting them in cold blood. Seemingly, however, they preferred to execute them by hanging. They took them to the woods, tied a cord about the neck of each condemned feline in turn and attached the other end of the cord to the top of a springy sapling that they had bent over. When they released the sapling, it snapped back to the upright position, hanging the cat as effectively as from a gibbet.

Our cats were given a pan of milk, warm from the cow, at milking time. They looked forward to this in happy anticipation and were always at hand to lap up all in the pan. There was a wise old tom who would sit, purring contentedly in a corner, as a sort of self-satisfied supervisor when we were milking. He learned the trick of opening his mouth widely as an invitation for someone to squirt it full of milk. He never minded in the least if, due to incorrect aim, the milk spattered over his face; he just swabbed it well with a paw, licked the paw, and got set for another shot.

I was a toddler, too young to remember the first dog that came into my life; but I heard the story told a good many times. My parents had come home from an afternoon visit somewhere. Just as they lifted me from the carriage and set me on the ground, Dad's bulldog, an animal of which he was very fond and that up to that time had been perfectly gentle and friendly, suddenly set his teeth into my throat (the scars have been there ever since). Seizing a stick of stovewood, Dad slew the dog on the spot.

Penny was a rather large dog, gentle and safe as a playmate for small boys. He was intelligent, alert, a good stock dog, a marvelous rat catcher, and as fleet of foot as a greyhound. We were strolling through a field one day, when Penny's sharp eyes fell upon a weasel running along the rails of a "snake" fence. A few long leaps brought him abreast of the weasel, and, when it sprang across a corner, Penny seized it in mid-air. His sharp teeth ended in a twinkling the career of that destructive animal. For weeks afterward the dog was embarrassed by a skunk-like odor that clung to him—in the momentary encounter the weasel had got in a shot from its artillery.

After he had been with us five or six years, Penny fell into the evil habit of prowling away from home at night. The habit grew until at last he spent nearly all of the daylight hours, when he was at home, sleeping. Thus he became practically worthless. Dad was very suspicious of the dog's nocturnal wanderings, feeling certain that if he were not already a sheep-killer he soon would become one. Prowling dogs often gather in packs. The marauding habits of their wild ancestors then tend to assert themselves. They begin to chase sheep, at first in fun. If one dies or is killed, they tear and rend the carcass, eating parts of it. The excitement and the taste of blood confirm them as killers. There was then, as now, a county fund from which owners were paid compensation for sheep killed by dogs. Nevertheless, farmers regarded sheep-killing dogs with bitter dislike, and the man who harbored one sooner or later found himself in bad grace with his neighbors. To be on the safe side, Dad took Penny away and had him shot. Very early the following morning, after a night of rain, a man who lived on a farm four or five miles away knocked at our door. "Bill," he said, "that dog of yours was chasing my sheep last night. Just tracked him in the mud right here to your house." When Dad protested that our dog had been shot the day before, the man stammered shame-

111

facedly that some mistake must have been made. We were all sure of that, and were glad that we hadn't made the mistake of keeping Penny a single day longer.

The smartest and best dog we ever had was Kelly, a small Scotch collie. My brother bought him when he was a tiny pup. The first night he was with us was a painfully wild and restless one. No sooner did we get to bed than that homesick pup set up the most heart-rending wailing and yelping imaginable. Never, I suspect, has canine heart-break been so vociferously articulate. There was no comforting him; nothing that could be done served to assuage his sorrow or quiet his distressful cries. Much sharp criticism and many ill-natured looks were directed at the dog's master but, undaunted and undismayed, he stood up loyally for his grieving charge. Gradually, as the days wore on, the dog became accustomed to his new home and was able to sleep quietly at night. He early developed an astonishing talent for wallowing down and uprooting Mother's flowers; but, luckily for all concerned, he soon gave that up for the exciting sport of stalking crickets. Next, he acquired the art of digging out and killing mice; he became a relentless foe of rats as well. He learned to be very useful in driving livestock. He was an exceptionally alert watchdog, loved to join in rough tussles, and was unexcelled at hunting rabbits.

The first independent carpenter work we boys tackled was the construction of a dog house. We wasted some good material and displayed a shocking lack of skill, but we finally got the thing assembled and nailed together. There was no denying that, from an artistic and architectural standpoint, it left a great deal to be desired; yet none of our dogs seemed to mind its deficiencies much or

112

to find other ground for serious criticism. With plenty of straw on the floor, they curled up in it and spent their nights in comfortable slumber.

Once an aunt gave us a pair of guineas. Wild and skittish birds, we had a hard time getting used to them; no doubt the difficulty was mutual. Their sharp eyes, like those of plane-spotters of early cold-war days, seemed ever to be scanning the skies. At the sight of any kind of hawk or other big bird they would set up a series of dreadfully raucous squawks—"Pot-rack! Pot-rack! Pot-rack!" If these cries didn't actually scare away the hawks, they at least served to warn all the other fowl on the place to take cover. They were powerful fliers; soaring high above the tall barn was child's play for them. They seemed to derive enormous satisfaction from perching at the very pinnacle of the roof. We gathered two dark-shelled eggs from the guineas' nests every day. We enjoyed no end eating them— our very own eggs. We liked especially to take them, hard-boiled, in school lunches; they had very hard shells and it was fun to crack them against the heads of unsuspecting schoolmates. In a short time we found that there were more eggs than we could eat. Why not, we asked our-selves, set the eggs and go into the guinea business in a big way? Accordingly, we selected what we considered a trustworthy hen chicken with a strong determination to do some incubating. A full clutch of eggs was placed under her and we withdrew to let nature take its course. Three weeks went by, and not an egg pipped. The fourth week passed without results. Vainly, we waited and watched a few more days. The patient hen was all for going ahead with the venture, but we were convinced that proceeding further with those eggs would be a sheer waste of time for all parties concerned. We now began putting the obvious two and two together: since two guineas were laying two eggs per day, and not an egg had hatched, both guineas, we concluded, just had to be hens.

We raised turkeys several summers, but we never liked them (on the hoof). During their first few weeks of life, they are so delicate and so easily killed by chilling or by disease that they require a great deal of care. No matter how good the roosts and shelters we provided might be, with mulelike stubbornness they invariably ignored them completely and flew high into the trees for their nightly repose. We might have excused that if they could have been content to remain at home by day; they never were. Our turkeys preferred to wander all summer long over neighboring farms, often a mile or two distant, while our neighbors' turkeys foraged over our farm. Because of their roaming proclivities we always marked them for identification by tying to one leg of each a strip of cloth of a certain color. When time came to start feeding them for market in the fall, we had to round them up like cattle on the range, drive home the birds wearing our bands and then contrive somehow to keep them there.

Late one summer, we completely lost track of our turkeys. We hunted far and near and made diligent inquiries but no trace of them could be found. Since the flock of 100, almost fully grown, was worth some real money, I set out to make a thorough search. I came upon a flock shut up in a pen on a farm some two miles from home. I noted that they were of a size and color to be ours; the number tallied exactly, and tied to the right leg of each bird was a yellow cloth band (all of our turkeys had been marked that way). At the house I suggested as

114

tactfully as I could to the young woman who answered my knock that our turkeys seemed somehow to have gotten into their pen. "Oh, no," she said, a little confusedly, "they're not yours. Can't be. Why, we raised 'em, every one. I'll swear on a stack of Bibles as high as this house that they're ours." I mentioned the yellow cloth markers, the size and number of the birds, their color—all tending, I suggested, to identify them as ours. All that, she declared, proved nothing. Approximately a week later those turkeys, not one missing, came strolling into our feed lot, and there they remained until we caught and crated them for the Thanksgiving market.

We got along better with chickens than with turkeys; they rarely strayed far from the feed lot. A little coop was provided to shelter each mother hen and her brood. Each morning in nice weather each little family was released from its coop to wander at will. A brick was set on end under a corner of the coop so that the hen and chicks could get back at nightfall. As a rule, the coops were let down after all were safely inside.

One evening this was forgotten. Toward midnight the household was awakened by a hen's terrified squawking and a loud commotion that seemed to come from one of those little coops. Thinking of chicken thieves, we all ran to our windows and peered out in the direction of the sound. It was so dark that nothing could be seen. The hen's loud cries continuing, Dad fired a shot from his revolver. Immediately there was a resounding crash; then all was still. We went to the coop at daybreak and were astonished to find in one corner a giant horned owl; in the opposite corner, as far from the owl as she could possibly push herself, was the hen. Peeping dejectedly

115

at a safe distance outside were the chicks. The sequence of the night's events was clear; the owl, inside the coop, had been so startled by the pistol shot that he knocked over the brick and the coop fell, making him a prisoner. One can well imagine the terror of the night for the hen, but she came through unscathed and all the chicks were safe. His hands protected by heavy mittens—probably not needed at all because not a particle of fighting spirit remained in the owl—Dad pulled the marauder from the coop and caged him. He was an unusually fine, large specimen; but no one in town could be interested in taking him to be mounted for display purposes, so Dad carried him to the chopping block and beheaded him.

A few hens in every flock of chickens would hide their nests where no one could get at them. It seemed to afford them inordinate pleasure to stroll out nonchalantly in the fullness of time with a big brood of chicks. There was no particular objection to this, except that in about half of all such cases they brought forth their progeny very late in the fall; then the chicks, scantily feathered, might not survive the low temperatures from that time on.

ELEVEN

The Beginning of the Farmer's Year

IF THE WEATHER PERMITTED, the year's farming began in March with the sowing of oats. One year we got this job done in late February and were rewarded by an exceptionally large yield of excellent quality grain. Often, oats went into a field where corn had grown the year before. If there was time, the ground was plowed; but in most cases it was broken up by going over it once or twice with a spring-tooth harrow, followed by a spike-tooth harrow, to obtain a deep, well-pulverized seedbed. A disk harrow and a cultipacker would have been very helpful in this work, but neither was available until several years later.

We used a grain drill for putting in oats and other small grains. Nearly always, the drill was set to feed out with the grain seed a regulated quantity of grass seed. Red clover was commonly sown with oats and timothy with wheat. No grain drill in the area at that time had an attachment for applying fertilizers—commercial fertilizers were generally unknown until about 1919, when a sample was shown at the county fair. All seed grains were carefully selected. We always ran ours through a fanning mill to remove weed seeds, chaff, and bits of straw. Clover seed was sown over wheat fields in early spring by means of a "fiddle bow" broadcast seeder. Soon afterward, the field was rolled.

Near the time we finished with the oats, we started plowing fields for corn. It was impossible to hurry any spring work at the outset because, having done little work through the winter, horses were "soft." Frequent rests were imperative when the weather was warm. One perfect spring day, about mid-forenoon, I stopped my panting, sweaty horses to let them rest and blow. I stretched out on the warm ground beside the plow for a little breather myself. When I opened my eyes after an unintentional nap, the horses were still standing quietly, half-asleep. It was high noon—dinner time. (The noon meal was always dinner and the evening meal supper.) That little snooze could have caused me a great deal of embarrassment but, taught by previous experience, I kept it a secret.

With steady, well-trained horses and a plow sharpened and adjusted properly, plowing was not altogether an unpleasant task. The hard work came at the corners, in getting the plow out and then back into the ground for the change in direction. Although the pace was slow, walking a full day in the furrow in crumbly, sticky soil was fatiguing. Operating a plow in stony soil or in stumpy new ground constantly exercised about every muscle in one's body; he had to remain tense and alert to avoid being violently jerked or thrown in case the plow struck an unyielding obstruction. Plows we used were of the single-bottom, walking type, now rarely seen. The plowman had plenty of company all day long; flocks of robins, chickadees, sparrows, jays, pigeons, catbirds, and blackbirds followed just behind him to eat worms, bugs, and insect larvae as they were turned out. Occasionally, as in

Robert Burns's experience, a frightened mouse was unearthed, running in panic from its wrecked home.

Soil that was loose after being turned by the plow could be readily fitted for planting by means of a spike-tooth harrow. In case heavy rains after plowing caused the soil to "run together," it had to be first broken up with a spring-tooth harrow. At times, after plowing stiff sod fields, we had to use a heavy roller to press down the long strips of broken turf. Whether the roller was used or not, we gave broken-sod fields a thorough going over with a spring-tooth harrow, following up with a spike-tooth. Generally, the final fitting was done with a slant-tooth harrow that, the last time over the field, was dragged diagonally so that marks later made to guide planters would stand out distinctly.

There was a tradition, attributed to the Indians and fairly well substantiated by the white man's experience, that corn should be planted when the leaves of the white oak were as big as a squirrel's ear. Generally, this was about May 10. The first step in planting was to lay out the field in rows. This was done with a homemade sled-like implement that marked on the ground three rows each time it was driven across the field. Standing upon it, one guided the horses in a straight line by sighting at tall stakes set and reset for the purpose. After the field had been marked in one direction, another set of marks was made at right angles to the first so that the entire area was finally laid out in squares measuring about forty inches on a side. A hill was to be planted at each corner of the squares.

About the time the marker started on the cross rows, work with hand "jobber" planters began. At the bottom

119

of the planter's seed hopper was a small wheel that had several holes of uniform size near its outer edge. Each hole was supposed to hold three grains, considered the proper number for a hill. Using this planter involved a series of rhythmic movements. As you set your left foot beside a spot marked for a hill, you stabbed in the closed bottom blades, with three grains waiting, and at once brought the two handles together. That dropped the seeds and the soil closed over them. Moving right along, you pulled the handles apart as your right foot advanced. That closed the blades again, rotated the wheel one notch, and dropped three grains into the pocket formed by the blades, ready for the next hill. The work was fun for a while, but it soon became a monotonous grind. The necessary walking posture and handling the planter were fatiguing, particularly to one's back. Multiple-row planting machines now do this work and do it well.

At once after the corn shoots appeared above ground, we went over our fields with a slant-tooth harrow, if the soil was not too wet. This gave us the jump on the weeds, at that time tiny and tender, destroying them by the million. Thereafter, we kept the single-row riding cultivator going. At times we "speeded up" the work by using also a one-horse, double-shovel walking cultivator that did one row per round. Farmers generally tried to get over their corn with the cultivator at least four times at weekly intervals; often, rains and the necessity for harvesting hay upset such plans.

On a good many farms some space was reserved in a cornfield for a melon patch. A few times we located ours near the center of a field, hopefully believing that it would thus be concealed from predatory youngsters

120

(and oldsters) of the neighborhood. The stratagem was never successful; the patch was always spotted by prospective raiders long before the corn grew tall enough to hide it. Our luck was a lot better with patches located nearer the house.

A number of times we put in with our corn a long row of broomcorn. We used the long, tough stalks for tying corn shocks. One fall we chopped off the bushy tops, stripped away the seeds, and took the straw to a broommaker, who turned out enough good brooms to supply us for several years.

At that time a good many farmers reserved a plot for sorghum cane. When ripe, the tops were clipped off and all leaves were stripped away. Then, the stalks were cut off close to the ground and tied in bundles. At the mill the stalks were fed between rollers and crushed to a dry, stringy pulp. The juice flowing out was caught in a receptacle and transferred to evaporating pans. There it was boiled, and skimmed from time to time, until it became a moderately thick syrup. Housewives used the syrup in baking and as a sugar substitute in other cookery. It was then an everyday commodity that commanded only a nominal price; now, if one can find any, the price runs to several dollars per gallon.

Early each summer, hordes of striped beetles appeared and began to devour the foliage of our potato plants. They were similar in size and appearance to the lightning bug or firefly indigenous to the area. We boys called them "blister bugs" because if we happened to crush one in contact with the skin, a blister would appear there within a short time. No one knew any way to combat them except to rout them from the plants and then, as they scurried over the ground, whip them vigorously with brushy branches from a tree. Crude though it was, this procedure was successful—at least temporarily. The pests left, crawling and flying away in great haste, almost with an air of remorse and penitence. We often wondered where they went, suspecting that they very shortly would show up in a neighbor's patch. No doubt when he whipped them, they hurried right back to take up where they had left off in our potatoes.

Once Dad found a heavy infestation of these beetles in our potatoes and sent us boys to whip them out. While glumly girding for the fray, an inspiration came—why not let our ducks do the work? We promptly drove in the flock, some fifty or sixty half-mature birds that belonged to my brother and me. It was heart-warming to see our white Pekings go after those bugs. They waddled down the line gobbling the pests right and left until, within an hour, the last one had disappeared. Then, noting the distended crops and recalling how our bare knees had once been blistered by bugs we crushed as we crawled along plying our whips, the alarming question arose: "What dreadful thing is going to happen to those beetle-stuffed ducks?" Luckily, nothing happened at all; they went back to their pen, serene and satisfied, and no symptoms of any kind of trouble were noted.

One year the long beetles did not come in bothersome numbers. Instead, we found on the potatoes a few squat beetles with wing cases striped in black and yellow. Though we didn't know it then, these were Colorado potato beetles, which have long made trouble for potato growers in many areas. We brushed all we found on the plants into a pail containing a small amount of kerosene.

122

This was easy because, possum-like, the adults feign death when they are touched or jarred and drop like stones from the plants. The oil no doubt would have killed them, but we made this doubly sure by burning them in the oil.

The actual effect of this procedure, so far as protecting the potatoes was concerned, was practically nil. The groundwork for its failure was laid when the first female adults, extraordinarily prolific creatures, began depositing yellow egg masses on the underside of the first leaves that appeared. After a short incubation period, small, soft-bodied larvae came from the eggs. Their total number was legion. They were born with insatiable appetites, and immediately began feeding upon the foliage. It is at this stage that the pests cause practically all of the damage. Reaching adulthood within a short time, a new reproductive cycle begins. Unlike mature beetles, the larvae cling so tenaciously to leaves and stems that it is almost impossible to dislodge them. We held them in check by applying arsenate of lead or other poison to the plants.

In the fall potatoes for winter use were piled conically on a layer of straw in the garden. We spread over them a blanket of straw, then shoveled earth over the straw, finally placing a piece of tough sod at the top to reduce washing down of the soil. Apples, beets, turnips, and cabbage were stored the same way, except that only potatoes and apples got a straw covering.

TWELVE

Farmers' Payday

EARLY IN JULY, WE HAD TO "lay by" the corn and start the hay harvest. On a day that seemed to promise favorable weather, we started the mowing machine. We tried to keep the amount of grass cut each day within proper limits, to avoid rain damage to dry hay that we might have to leave on the ground too long. After the grass had wilted, we went over it with the tedder, which turned and stirred it, opening it to sun and air two mower-swaths at a time. Next, the dry hay was raked into windrows. We used at first a wooden contraption with teeth set into a beam that made a half-revolution when the operator, walking behind it, lifted the handle. This deposited the accumulated hay in a long heap, as part of a windrow, and automatically positioned the second set of teeth for further raking. After a few years we put aside the wooden rake and bought a new one, all steel, with springy curved teeth. Like the wooden rake, it was drawn by one horse; but it was mounted on wheels, and it had a seat for the operator. The hay that accumulated as it moved along was deposited as part of a windrow when a hand lever was lifted. In heavy hay the fellow riding the rake might have had worries, but unemployment would not have been one of them. On rough ground, occupying that seat was almost equivalent to riding a bucking bronco.

At an early age both of us boys were assigned jobs in taking up hay from windrows, which often had been made by one of us. One drove the team hitched to the wagon on which was mounted a wooden rack about eight feet wide and sixteen feet long, made by our blacksmith. The other, with a fork, opened windrows for the passage of the wagon and piled up little cocks to aid the pitchers. A man on the wagon put the big wads of hay in place as the pitchers, with long-handled, three-tined forks, raised them from the ground. It was his responsibility to keep the hay properly spread and balanced to avoid tipping. The upsetting of a load could be extremely funny to a bystander; to the men who had to get the hay back into place, however, it was anything but funny.

Pitching from windrows required a lot of muscle and plenty of stamina—it was no job for a pantywaist. The pitcher on one side was supposed to throw up as much hay as the fellow on the other side, even if a wind was aiding one and hindering the other. The job became increasingly harder because the load grew steadily higher.

In time a mechanical loader came upon the market. Many who had an intimate acquaintance with pitching hailed it with delight, but their enthusiasm tended to wane rapidly when they got a sweaty workout with the machine. The loader, drawn at the rear of the wagon, astride a windrow, gathered the hay and rolled it onto the rack in great tangled masses. Two men found it a toilsome job to dispose of it as fast as it came up and build it into a stable load. Work imposed upon the horses was doubled but there was a marked saving of time.

126

Coming not long after the loader was a side-delivery rake that automatically deposited the hay in a continuous windrow, in spiral form, about the field. This was a decided advantage in using the loader. Most hay now is picked up from mower swaths or windrows by a machine that compresses it into bales as it moves along.

Each load of hay was driven onto the threshing floor at the barn. There, a man thrust a big two-pronged fork deeply into the hay. He then lifted a pair of levers; this thrust out fingers at the lower ends of the prongs that, harpoon-fashion, held the hay on the fork. The forkful, at times half a ton, was hoisted vertically by a rope running over a system of pulleys, the motive power being furnished by a horse hitched to the end of the rope and led or ridden by a boy outside the barn. Directly above the wagon a special pulley block attached to the harpoon fork contacted a trigger on a waiting car mounted on the track just under the peak of the roof. Mechanism in the car seized the block, and the car with its big wad of hay rolled to the right or left, depending on the roping. A tug on a light rope attached to the fork levers retracted the prong fingers and the hay fell, to be distributed by men working in the mow. In some barns the hay was dropped upon a pole laid across the mow, intended to throw it to one side or the other, to reduce work for men in the mow; we found this to have no advantage.

Nearly always, we had some hay left over at the end of the winter feeding season. If the amount was large, a baler was set up on the threshing floor and the hay was put into bales convenient for market handling. Now and then, a neighbor who had run short came with his wagon and bought an unbaled load. A few times we hauled loose hay to people in town who kept horses or cows. No one

127

enjoyed this because we always had to drive through some narrow, unpaved alley and pitch the hay through a small opening high in the wall of the customer's barn.

We had a set of slings made of ropes with wooden spreaders. Each sling was in two sections, provided with hoisting rings and a center catch that held the sections together. We used slings for hoisting bundled fodder or grain in sheaves into the mows, never for handling hay. In the field a sling was spread over the wagon rack. After a fourth of the load was in place, a second sling was spread, and so on, until all four slings were loaded. On the threshing floor at the barn a ring at either end of a sling was hooked to a pulley block let down on a rope from the car above. The sling with its burden was then hoisted the same as a harpoonload of hay. Pulling a trip rope released the center catch; the two sections of the sling separated, and the load dropped into the mow.

Wheat harvest came soon after haying; soon afterward came the oats. There was so much pressing work in the harvest season that farm boys didn't always get to attend picnics or other Fourth of July doings. A binder (so called because, after cutting off stalks of standing grain, it bound them into sheaves with twine) was used for harvesting all small grains. Three horses were needed to pull the heavy machine, whose mechanism was driven by a "bull wheel" in contact with the ground. McCormick binders were used extensively; other well known names among manufacturers of harvesting machinery were Deering and Walter A. Wood.

Operating a binder required several aptitudes and skills; it was a job almost as complicated as piloting a big plane. One had to control and guide the horses, all much more

128

interested in a chance to nip off grain to eat than in getting it harvested. Care was necessary to avoid over-heating the animals. The operator had to look out for possible obstructions hidden by the tall grain. He had to see that all parts of the machine were working properly. Oil had to be supplied to important bearings from time to time, and the twine supply had to be replenished as necessary. In addition to all this one had to release the carrier to drop its burden of sheaves at frequent intervals and then return it quickly to the receiving position. (For several years binders were made without bundle-carriers. Often it was the task of a boy or two to gather the sheaves as dropped singly by the machine and pile them con-veniently for men doing the shocking.)

In putting sheaves into shocks, the harsh, stiff straw and the beards scratched hands and arms severely. Thistles, which could prick fingers cruelly, might be bound in some of the sheaves. Much carrying and lifting were involved and walking through stubbles, often long and tangled, was extremely wearisome. In a big field the binder would get far ahead of the shockers, but as the area of standing grain was whittled smaller and a larger amount of time was required in turning at corners, the shockers generally began to catch up. Under average conditions two good men could finish shocking at about the time the binder got through, provided that they were stouthearted men who could withstand the temptation to give an undue amount of time to drinking from the water jug and resting in the shade of a tree.

A man who worked for us one summer told me that he was once shocking alone for a farmer known as a driver of his help. "I was putting the best I had into the job," he said, "but the binder soon got far ahead of me. Later, as the rounds got shorter for the machine, I seemed to gain a little. Noting this, the boss repeatedly urged me to speed up. He seemed to be determined that when the last sheaf dropped I should be standing by to put it into the last shock. Then and there I became even more determined that I should not."

For farm boys the chief thrill at hay and grain harvest time came when the mower or binder was nearing the last swath remaining in the field. Always at that time, a scared rabbit or two would break from the scant remaining cover and dash for a safer hiding place, boys and dog in excited, but generally futile, pursuit. Quail, with a deeply implanted instinct for the protection of their young, always managed to lead their babies safely out in good time. Occasionally, a wheel would run smack through a nest full of eggs, but more often neither nest nor eggs were damaged. The end result, though, was the same in either case; no quail would return to a nest once it had been exposed. Cats hunting in fields at harvesttime often had narrow escapes from tragic accidents. None of ours got hurt, but cats have been known to go through a considerable part of their nine lives on three legs, the fourth having been amputated by the blades of a mowing machine. Hens with nests in hay fields were occasionally cut to pieces by the mower.

Before mechanical harvesting machines were built, the grain cradle, which first came into use about the time of the War of 1812, was the only implement other than the sickle available for cutting grain crops. As a matter of fact, their fields were so stumpy that no early farmers could have used machines if they had had them. On our farm the cradle was used occasionally to cut grain in small enclosed spaces or in corners where the binder could not reach.

A neighbor told me that when the first binder appeared in the community where he formerly lived, a group of farm laborers one night piled sheaves about it and burned it to a mass of useless junk. Those men, of course, were under the delusion now common among men who depend for their daily bread upon the labors of their hands in industry: that the new machine, capable of saving an enormous amount of time and labor, would deprive them of an important part of their livelihood.

No such demonstration against the combine was reported when it came upon the scene some decades after the binder. There might well have been some violent reaction, however, from men who tend to view with an unfriendly eye supposed encroachments of mechanical innovations designed to render unnecessary the work of human hands. As its name indicates, this machine combines in a single unit mechanisms for performing the functions of both harvester and thresher. It eliminates the handling of many tons of straw and saves many man-hours of labor.

Threshing machines, also called separators, were powered by traction-type steam engines. To many farm boys those engines were among the most marvelous and fascinating of all machines. One sees them take in such everyday materials as wood, or coal, and water and transform stored-up energies into smooth, useful power. In full view are the spinning flywheel, the piston rod flashing in and out of the cylinder at high speed, the slide block, the governor, and other working parts, all functioning with velvet smoothness, all perfectly co-ordinated, all under complete control.

The threshing outfit traveled caravan-like from job to job, the engine pulling the separator at about three miles per hour. Behind came a team of horses drawing a tank of water on wheels and a supply of fuel. At one time there was a fourth unit in the procession, a contraption called a "katydid." Its purpose was to carry straw away from the stubby conveyor then built into the rear end of the thresher. When new machines came out with "cyclone stackers" the old separator and the katydid quickly disappeared. The new thresher had a big fan that ran at a fairly high speed inside a housing at the rear end. The fan forced the straw through a telescoping tube with considerable velocity. The discharge of straw could be directed at will by means of a rope-controlled deflecting hood at the upper end of the tube.

Each farmer supplied the fuel for his threshing. Often this was a pile of old half-length fence rails. The engineer, with plenty of time for the job, never minded chopping them into lengths that he could feed into the firebox; for some of the larger engines, little cutting had to be done.

Two men in addition to the engineer and the boy in charge of the water tank made up the crew of a threshing outfit. The two took turns at feeding the separator. Standing on a platform at either side of the feeder were two men, each with a knife in his hand. It was their job to cut the twine bands of the sheaves as they were deposited on the feeding tables, then push them toward the feeder who, with a deft twist, spread the straw evenly within reach of the high-speed cylinder whose steel spikes, run-

132

ning closely between similar spikes in the concaves, beat out the grain. Movements of the band-cutters had to be watchfully co-ordinated with those of the feeder in order to avoid cutting the latter's fingers. After the blower stacker, the next big improvement in separators was a mechanical self-feeding attachment. This did the work of the human feeder and the two band-cutters.

In the older machines the threshed grain was delivered from a spout at one side. The man assigned to measure the grain leveled a shallow measuring box beneath the spout; he had two half-bushel measures. When one, inside the box, was filled he pulled it toward him and at the same time slipped the other in place under the spout. Each time a measure was drawn over the floor of the box, a trigger caused the tally mechanism to register. As one measure filled, the other was emptied into a bag, held open by a boy. Stroke measuring, as this was called, was no job for a man with a weak back or a streak of laziness.

About the time the self-feeder appeared, automatic weighing attachments came into general use on separators. These ended stroke measuring. Threshed grain was conducted into the weigher, which could be adjusted to trip at the weight of a bushel or half bushel of the grain being handled. From the weigher a tube carried the grain either to a bagging device or directly to a wagon box.

Each outfit was owned by a farmer-operator or by a partnership of farmers. One machine commonly did all the threshing in the area about its home base, and that provided about all the work that could be done in a season. Threshers generally were pleasant, jovial men, all hustlers, constantly driving ahead to get all possible work done while weather conditions were favorable. All were mechanically-minded—no one else had any business monkeying with those machines. In addition to grain separators, several owned clover hullers, corn huskers, and portable sawmills; thus they got extra service from their engines. Few made much real money in the threshing business. Most equipment was so well built that with reasonable care it could have been used many seasons, but important improvements generally came so fast that separators became obsolete long before they were actually worn out. Owners had to turn them in for new ones or see the bulk of the business go to competitors with more up-to-date outfits. These trades required big outlays of cash, for all such machinery was expensive.

Outstanding among the threshers I knew was a man of exceptionally fine character who, no matter how provocative a balky machine might become, no matter how irritating conditions might be, never lost his temper, never got excited, never resorted to profanity. Often, when summer days were hot, he used an expression borrowed from comrades in the Union army: "Six weeks in the hot month of August and nothing to drink but water!"

All work of threshing other than that done by the machine crew was on a co-operative basis. Each man, planning ahead, would determine just how many helpers would be needed. Then he would invite this number of neighbors to help him, all of them understanding that he would return the favor in due time. Jugs of fresh drinking water were kept close at hand from morning until night— the water boy was frequently the busiest hand on the job.

The threshing dinner was traditionally a bounteous feast. Usually the *pièce de résistance* was prime roast beef, with great quantities of rich noodles and potatoes.

134

There were garden vegetables in abundance and fruits, relishes, and desserts in considerable variety. Tables were set in the dining room or a big kitchen with places for about two dozen men. Before eating, the men, some extremely dirty, washed hands and faces in cold well water in basins and tubs outdoors. Since there was always a great hurry and bustle, this washing tended to be quite sketchy—most of the grime was wiped off on the towels. When those at the first table finished eating, they hurried out; the women quickly put down clean table service, and the places filled with a second group of hungry men. The speed with which most of those fellows gulped huge meals, washing them down with water and coffee, was astounding. No doubt all would soon have become hopeless dyspeptics if they had continued long beyond the threshing season to eat so much so fast.

The women began a day or two in advance to plan and prepare those dinners. If rain stopped threshing, even for one day, and thus set back the schedule, it was almost a major calamity for them: because refrigeration was lacking, a lot of food was sure to be spoiled, and much of the hard work they had done would go for naught. Like the men, they had plenty of help, having invited friends and neighbors to aid them.

Some years before the combine came into use, groups of farmers in some instances organized what they called "threshing rings," pooled their capital, and bought out individual owners of threshing outfits. Ordinarily, a group included enough members to provide all needed manpower and to assure a grain acreage large enough to keep the equipment busy during a normal season. Competent men were chosen to serve as crew members. In

135

each case their duties were the same as those of a crew of a privately owned rig, and they remained at their posts until all of the group's threshing was done. Each member had a voice in determining the yearly threshing itinerary. Accountings were made at the end of each season, and members shared profits or losses resulting from threshing, at fixed rates per bushel.

About a quarter century has passed since the steam engine and the threshing machine were crowded into retirement, but they remain as cherished memories to hundreds of men who once worked with them. A surprising number of old engines and not a few separators have been retained by their original owners or, in some cases, by enthusiasts of a later generation. The engines are given special care; new paint is applied as needed and working parts are kept in perfect adjustment. In some instances owners install apparatus formerly undreamed of, such as generators to supply current for electric headlights, rubber-tired front wheels, and weatherproof cabs. A goodly number of men of the old school belong to threshers' organizations and hold annual get-togethers at central points. They move in, on low-slung trailers, a score or more of engines, all spick-and-span, and in perfect running order. They fire up, open throttles, and set flywheels spinning; they enter the machines in pulling and hill-climbing contests; they check and compare power output. Before the meet ends, they belt engines to well-preserved separators or sawmills and do a bit of threshing or sawing.

Not long ago a group of Black Swamp farmers persuaded one of their number to overhaul and tune up his binder, idle for many years, and arranged to have it used to harvest a field of oats. When the shocked grain was dry, a separator and a steam engine, both long on the retired list, were set up. Neighbors went in with forks and tractor-drawn wagons to haul sheaves from field shocks and feed them into the separator. For all participants this threshing was not work but an exciting, happy lark. There had been only grapevine announcements of the event, but several hundred people gathered from farms and towns in the area to watch the goings on.

After the threshing season was over, we generally found some time for the monotonous, thankless task of battling weeds. All summer long, as time allowed, we attacked them with scythe, shovel, or hoe. We chopped away at thistles, docks, mulleins, ragweeds, ironweeds, and a host of other weeds, kept them down, and allowed few to go to seed. (We even treated rye as a weed if any grew in the wheat.) Still, their ranks never dwindled much from year to year. A big patch of dogfennel appeared every year in the orchard. We mowed the weeds at least once each summer, but this had no permanent effect; the following year they came up as thick and vigorous as ever. The dogfennel—it had several other names, including May-weed and heath aster—may or may not have had the valuable medicinal properties ascribed to it by a few old grandmothers; to us it was just a nuisance. No animal would eat it; it had an unpleasant odor; it was unsightly; and it encumbered the ground.

It is said that Johnny Appleseed (Jonathan Chapman, an eccentric character of frontier days who spent the greater part of his life journeying to-and-fro in areas to the west of Pennsylvania, planting apple seeds as he went, and maintaining apple nurseries at several points) was

137

responsible for introducing dogfennel into Ohio. Believing the plant to be useful in the treatment of malaria, he collected seeds in Pennsylvania, carried them westward, and planted a few in the dooryard of every home he visited. We had no knowledge that Johnny ever came into our immediate neighborhood, but it is altogether possible that our venerable apple trees came more or less directly from one of his wilderness nurseries—also, that our never-failing dogfennel patch was a result of his kindly meant effort.

We felt that the number of weed species we had to contend with was more than large enough. It is even larger now. Among agencies that may be suspected as responsible for introducing seeds of additional weed pests —accidentally, for the most part—are machines moved from farm to farm, winds, railroad equipment, and commercial trucks. No doubt some were introduced in commercial grass or field seeds. It may even be, as some believe, that a few have traveled hundreds of miles to the region in dust storms. Probably the worst of the weeds that are now major nuisances on Black Swamp farms is the highly prolific Canada thistle. It grows in large patches on many farms, in some instances having taken over entire fields, and it is spreading rapidly.

THIRTEEN

The Fall: Full Job Quotas

corn and set it up about "galluses" to form shocks that
they might classify as "ten hills square" or "twelve hills
square." The work was all done by hand, the stalks being
cut off with a knife essentially the same as the all-purpose
machete used in jungle lands. This was one of the hardest
jobs on the farm. It involved lugging heavy loads of ear-
burdened stalks, quite often over ground that, due to
tangled weeds, could hardly have been more difficult for
walking. Moreover, those September days, more often than
not, were about the hottest of the year. Only by cutting
and shocking could corn fodder be preserved in good
condition for use as winter forage for livestock. (On the
farm the word "fodder" was not used in the broad, general sense to include all coarse food for domestic animals;
ordinarily, it was applied only to the mature, dry corn
plant, usually minus the ears.) Another reason for putting
the corn into shocks was to clear the ground so that wheat
could be sown in it.

Each fall, soon after the corn had been cut and shocked,
the ground was broken up with harrows and winter wheat
was drilled in. One spring we found large patches of dead

wheat plants; in other parts of the field the plants were pale and puny. We were aware that the Hessian fly (so named because it was believed to have been brought to America in straw used by Hessian soldiers who aided the British in the American Revolution) was the villain in the piece. This was the only time I ever saw fly damage to wheat on our farm. Farmers at that time knew that late sowing could prevent or reduce damage to wheat by this pest but they had no means of knowing just how long to hold off; no doubt we had sown too early that time. Beginning several years later, the Ohio Agricultural Experiment Station each fall has determined the safe sowing date—it comes after the fall generation of Hessian fly larvae has entered the "flaxseed" or pupal stage in which, usually, the insects remain dormant through the winter. The date, which varies with latitude and, from year to year, with weather and other local conditions, is relayed to wheat growers by the county agricultural agent.

Corn husking began as soon as the ears were dry enough to be stored safely, ordinarily in early October. This was a tedious, monotonous task because each stalk had to be handled individually; yet it could be moderately pleasant when the fodder was just damp enough to be soft and pliable and the ground remained dry. Opening and removing the husks were aided by a husking peg or a hook, attached by straps to one hand. (Some stripped away all silk and shreds of husks, figuring that these nest-building materials, if left on the ears, would encourage mice and rats to establish themselves in storage cribs. Men who set records for fast husking, however, never bothered about this.) As we husked, we piled the ears on the dry

142

spots where the shocks had stood. The fodder, tied in bundles, was set up in big shocks. The corn, later gathered in baskets and dumped into the wagon box, was hauled from the field. A broad, short-handled scoop shovel was used to toss it upward six or eight feet into the storage crib. After disposing of a load or two, the shoveler would have welcomed enthusiastically a mechanical elevator such as farmers now use, for shoveling was a fatiguing job that put a continuous strain upon muscles, particularly those of the back.

About the time husking began, we picked our supply of apples for winter. There were six or eight apple trees and one pear tree that had come with the farm, all with trunks ten to fifteen inches in diameter. Some were at least thirty feet high; no effort had ever been made to induce a low, spreading growth by pruning. Two or three of the trees had living crossties three or four inches in diameter, the ends of each having been grafted years before into lateral branches, one opposite the other. These grafts, intended to prevent spreading and breaking of the branches, were the only ones of the type I ever saw. In the bark of the huge branches of all those old outsize trees were numerous circles of round perforations made by sapsuckers and other species of woodpeckers in quest of burrowing larvae. Among the most numerous of all our birds at the time were the redheaded woodpeckers, more given to devouring fruit and the eggs of other birds than to foraging in bark for grubs. Though excessively noisy and quarrelsome—the ruffians of the family—they are very lively, interesting creatures. In recent years they have almost disappeared. Poisonous sprays may have had much to do with this, but I have been assured by a qualified

143

observer of bird life that the automobile has been the principal killer; on roads the birds fail to yield the right of way.

We had only two varieties of winter apples, Russets and Romanites (I have never seen the latter anywhere else). All of the apples were rather small because no fertilizer was applied about the trees and because thinning of the newly set fruit, as practiced by professional orchardists, was out of the question, due to the size of the trees. Although no spraying was done, the fruit was remarkably free from insect and disease damage.

After the picking was done, we shook from the trees a full wagonload of apples for cider. With enough barrels on top of the load to hold the cider, we would set off for the cider mill in town early the following morning. No matter how early our start, we always found ahead of us at the mill apple-filled wagons in a line a block or two long. They moved forward slowly, each driver in turn shoveling his apples into the hopper of the grinder, then waiting until his cider was pressed and transferred to his containers.

One barrel of our cider was reserved to be converted by natural processes into vinegar. We placed it, bung up, in a shady spot. With long straws thrust through the little air vent near the bung, we boys drank the sweet liquid by the quart. We were certain that there never was a beverage quite so delectable, chilled, as it was, to just the right temperature. Within two or three weeks fermentation made the cider unpalatable. Soon after that, we turned our attention to frozen apples knocked from the trees. The barrel with its remaining contents was transferred to the cellar where it "worked" and eventually became vinegar.

144

We had a neighbor or two who cared little for sweet cider; to them the stuff was not really potable until the fermentation process had gone far enough to give it a stiff wallop. They had their cider pressed late in the fall, kept it in a moderately warm place until it became "hard," and then rolled the barrels to an outside shed where it was cold enough to check further fermentation. They drank it throughout the winter.

For apple butter, cider was boiled in a large copper kettle until it was reduced to a fairly thick syrup—no one, it appears, ever thought of using anything but a copper or brass kettle for this purpose. A quantity of good cooking apples, peeled, cored, and quartered, was added to the syrup, the simmering contents of the kettle being stirred with a wooden paddle until cider and apples were perfectly blended. Ordinarily, no sugar or spice was added. The finished product had a piquant flavor and a reddish brown color, tending to darken with age.

Several walnut trees on the farm bore prolifically. The nuts began dropping off about mid-September; it was easy to shake off all that remained on the trees. We removed the pulpy hulls by pounding with a hammer as we held the nuts on a stone or block of wood, or we might drive them through a hole in a plank. Either way, it was a tedious job; and the juice that abounded in the pulp dyed our fingers a rich, deep brown. Some of our friends maintained that squashing ripe tomatoes and washing one's hands in the juice was the sovereign method for freeing them from the stain; others pooh-poohed this idea, declaring pontifically that the only really effective treatment was to insert the hands in a calf's mouth and allow the animal to suck at will. We tried both methods at one time or another; neither produced perfect results. We decided that the best thing we could do was to go about our regular business, and in a week or two our hands would regain their normal hue. After some experimenting one fall we found that we could save much time and avoid most of the stain by tightening the gears of the corn sheller and running the nuts through it. As they came from the machine, we spread them where they would dry in the sun. After thorough drying, separating the nuts from the shredded pulp was an easy matter.

Gathering hickory nuts involved a lot of pleasant tramping through leaf-strewn woods, resplendent in gorgeous autumn colors. Some alertness was required; if one failed to watch the hickories closely and get at them as soon as they were ready, someone was sure to beat him to the harvest. We could get the nuts off some of the trees by "bumping." To bump a tree, one selected a fairly heavy fence rail, walked ten or fifteen feet from the tree, poised the rail horizontally above his head, and ran full tilt to ram the end of the timber against the tree's boll. This would not work if the tree was a large one; in that case one had to climb high into its branches, shake vigorously, and beat off the nuts with a club. One sunny September afternoon a city-bred aunt, with several women friends, drove out from town to go hickory-nutting. We rustics had a hard time avoiding outright laughter at the expense of the city slickers; they had brought with them a coal scuttle and a small shovel, naïvely believing that harvesting the hickories was a simple matter of walking under the trees and scooping them from the ground.

Often we searched for beechnuts, which, though small, are exceptionally rich and delicious. Occasionally, we

146

found a few among the fallen leaves under a tree, and we might pluck some from the boughs. But with the best of luck we rarely gathered more than a handful or two; the canny squirrels beat us to them. We had walnuts and hickory nuts to nibble at all winter. Surplus nuts were sold to grocers in town. We might get two dollars for a bushel of good hickories. We were lucky if we got a dollar for a bushel of walnuts—product of the labor of two boys during about two full days.

If a cornfield bordered upon a wooded area, we could always be sure that many hills would be destroyed by squirrels. They would dig out and eat the newly planted seeds or pull out tender young shoots, still attached to the parent grains. They were also fond of eating at the ears when they were in the "milk" stage. Viewing the damage, we would vow determinedly that the score would be evened come squirrel-hunting time in the fall. Usually, we gave several days to the evening process, but without marked success. Squirrel stew was commonly regarded as a perfect food for invalids and convalescents. It happened occasionally that a local Nimrod, tired of humdrum work, would shoulder a gun, even in a very busy season, and take to the woods, excusing himself on the pretext that he wanted to shoot a squirrel for a sick friend.

For a long time there was no closed season on any game, but little hunting was done in the r-less months. Quail abounded, and many enjoyed shooting them. The main interest of farm boys was rabbits. They bagged them at every opportunity in fall and winter, tramping through fields and woods with dog and gun. Several times a week, the average lad would eat fried or roast rabbit

147

with gusto. A small culvert that ran under our driveway near the road was a favorite rabbit retreat. We looked into it frequently, and if one happened to be hiding there, one of us held a bag at one end of the duct while the other prodded the bunny out with a pole. Dressed carcasses were commonly hung outdoors overnight to freeze. This was supposed to remove the "wild" flavor and improve the meat.

The men of a family in our neighborhood always had one or more coon hounds. They made long trips occasionally to trade dogs or to buy replacements. Often in fall and winter, they invited parties of friends to join them in coon-hunting jaunts at night. All hands reported these hunts as providing capital fun and excitement. They netted little else, for this was quite a while before coon-skin coats had been found so essential on college campuses, and few of the hunters had any pronounced hankering for coon meat as an article of diet.

Although missing from the scene in our time, passenger pigeons, said to have been more numerous than any other birds in their far-flung range over the eastern United States, were at one time common in the Black Swamp region. It is recorded that migratory flights, which might continue for hours, almost shut out the light of the sun. They settled into trees in chosen roosting places in such numbers that big branches and even sizable trees broke under their weight. Men and boys slaughtered them on a wholesale scale by clubbing them as they slept. Hundreds were cooked and eaten, and surpluses were fed to hogs. Forays against them were everywhere so persistent that they rapidly dwindled in numbers. An individual believed to be the last survivor of the species died in the Cincinnati Zoo in 1914.

FOURTEEN

Work and Play for Farm Boys

AT ANY TIME OF THE YEAR
the amount of work that could be found to do on a farm
was practically unlimited. At times there were waiting
queues of tasks that simply had to be attended to within
definite time limits. This meant jobs for everyone, including even the children. They started early in life, with
simple chores suited to their years; as they grew older,
their responsibilities increased. Many boys after the age
of twelve became almost full-fledged "hands," occupied
almost every working day when there was no school. Generally, there was little or no compulsion about this; no
one was driven or held to rigid work programs. The custom
was so general and so well established that everybody,
including the young, took it as a matter of course.

A single exception comes to mind, a boy of about twelve.
Rebelling at what he considered an endless multiplicity
of tasks and a stern family discipline, he slipped away
from home one night, walked half a mile to a railroad,
and slogged off westward over the ties. The family missed
him, but no one worried unduly, assuming that sooner or
later he would be back. The second morning they found
him sleeping peacefully in his bed. Weariness, hunger,

and loneliness were factors that he had not taken into account at the outset.

As a rule, youthful workers were not paid regular wages, but occasionally they received cash gifts; allowances, as such, were unknown. In many cases they were encouraged to undertake side ventures to earn pocket money and to build savings accounts. All actual needs were of course supplied by parents. Opportunities for play and recreation abounded, and youngsters took full advantage of them. At least my brother and I did. A case in point was our leisurely handling of the task of berry-picking, about the time we reached school age. Gooseberries, blackberries, and raspberries grew wild along some of our fences. When they ripened, Mother sent the two of us from time to time to pick them. We combined the job with such activities as hunting birds' nests, shying stones and clods at all sorts of targets, exploring the creek for minnows and crawdads, hunting field mice, and lending an eager hand to our dog in chasing ground squirrels—"grinnies" to all local lads. Before the job was finished, we were pretty sure to lie a while in the shade of a tree, eating some of the berries, and abandoning ourselves to total idleness and relaxation.

Once we were sent out with a quart of climbing beans to be planted beside hills of corn in the field. For a while we pushed the seeds into the ground diligently; but the sun was hot, and our thoughts wandered to a variety of things to do that had much more appeal. Finally, yielding to temptation when the task was half completed, we dumped the unplanted beans into some tall grass and went with our dog for a ramble in the woods. It must have been incidents such as this that Dad had in mind when he re-

152

peated, from time to time, a homely old saying to the effect that one boy is one boy; two boys, half a boy; three boys, none at all.

Boys fared reasonably well so far as spending-money was concerned. They gathered and sold nuts, they marketed surplus popcorn and melons, and they accumulated rags, iron, and bones that brought a little cash from the junkman. One year my brother and I bought hatching eggs and raised a flock of ducks that made us some nice money—we had no overhead, and all of our feed was "for free." If they could be spared from work at home, boys at times went to help neighbors with field work. They were usually paid wages that were current for grown men, a dollar for a day of ten hours. No one then had thought of eight-hour days, forty-hour weeks, fringe benefits, or time-and-a-half or double-time for overtime.

For farm boys there were few idle hours and not too many dull ones. Often, the work they did was hard and tedious, but nearly always they managed to squeeze some fun out of doing it.

There was one job from which not even the most expert fun-finder could extract a single smidgen of entertainment or diversion: That was turning the grindstone, an extraordinarily tedious, monotonous task. Every farm had its grindstone mounted in a frame. It might be kept under a roof but in most cases it stood, month in and month out, somewhere under an apple tree. By means of a hand crank, some hapless lad had to keep the stone turning at a brisk rate and, at the same time, apply water to the cutting face while the grinding operation went on. A considerable number and variety of edge tools had to be sharpened in this manner—in depressed moments this num-

153

ber might seem astronomic. In time grindstones came mounted in frames that had a seat and a foot treadle with provision for applying water automatically. Many boys regarded that machine as almost equivalent, so far as they were concerned, to the Emancipation Proclamation, for with it their dads could do all necessary grinding without assistance. Later came high-geared grinders with emery or carborundum wheels that did in minutes work for which the old stones would have required hours.

None of us would have believed this at the time, but being kept busy and having a limited amount of money to spend was undoubtedly a good thing. Under such circumstances one learns to make good use of time; he tends to become resourceful, and he improvises to provide diversions. My brother and I collected thick pieces of bark from mature cottonwood trees, ideal material for all sorts of whittling. From some of it we carved boats that we launched and sailed in puddles after rains; lacking puddles, we sailed them in our watering trough "seas." Many a whistle did we fashion from willow twigs or pumpkin leaf stalks. We built small wagons and sleds, and a wooden engine that would run when we turned a crank. We even built a toy threshing machine that would thresh wheat (one head at a time) and a small baler that turned out miniature bales of hay. We made some of the tools needed for certain kinds of work—all a bit crude, but they worked.

One day I found a magazine advertisement offering a book, "Every Boy His Own Toy Maker." That, I decided, was just what I needed; so I mailed a dime, and back came a small paper-covered pamphlet containing instructions for making a variety of items. Our reaction to it was

154

rather negative because it seemed quite vague in spots, and it called for the use of several materials we had never so much as heard of, such as logwood, deal, teakwood, and draper's boxes. We did get from its pages instructions for making a telephone set, using as transmitter-receivers two tin cans with both ends cut out, a cord connecting the leather diaphragms, cut from an old shoe and tied over one end of each can. One really could talk over the contraption, but, considering our unaided vocal powers, it could scarcely have been regarded as a necessity.

We once built a kite that, for us, was of heroic size. We cut the frame from light basswood and pasted heavy wrapping paper over it. A long, heavy tail was attached, and binder twine was tied to the stout bridle cords to serve as the control string. It stood six feet high, and the way it flew would have delighted the heart of any boy.

Not all of our projects were successful. One time, for instance, we undertook to make a wooden bicycle, fastening the parts together with nails. Results, naturally, were far from satisfactory. Efforts to make a cash carrier to run over a wire track, like some we had seen in stores, likewise ended in failure.

From time to time we seined minnows from the creek and transferred them to the watering tank—nearly always, we had some there. Often, we speculated as to how long it would take them to grow to "whopper" size. Actually, none ever grew noticeably. We put into the water long hairs from the tails of horses, naïvely believing that, as some of our friends said, they would develop into living horsehair snakes. We always had a dog. It is very doubtful whether anything was ever invented that can equal a dog as a source of fun and entertainment for a boy on a farm; and dogs and farming fit together as if expressly made for each other.

Charlie, a lad several years older than we, spent much time at our house and, as a young man, made his home there for a few years. He was a genius at getting fun out of everything he did. It seemed to us that no one could equal him when it came to inventing exciting games and making toys. Probably no kids ever had more fun than we did when, with practically no effort, he induced us to chase him, in and out between buildings, over fences, around, under, and into trees. He made for us cornstalk fiddles and taught us how to play them with cornstalk bows. He cut out straight sections of elderberry stalks, punched out the central pith, and made us popguns (chewed paper wads served as ammunition). He taught us how to make leather slings with attached cords, essentially the same, no doubt, as the sling that David used in felling Goliath. We slew no giants, but we hurled countless small stones high into the air. A little later, we learned from schoolmates how to make slingshots, using wide rubber bands attached to wooden forks cut from trees. With these much greater accuracy was possible. Under Charlie's tutelage we cut slender, springy "water sprouts" from apple trees and whittled the smaller ends to sharp points. By impaling a green apple on such a stick and swinging it at high speed through a long arc, we could throw it with great force. One could not control these shots effectively, but the range pleased us no end. Charlie initiated us into the brotherhood of gun lovers when we were quite small by buying for each of us a long-barreled, wicked-looking cap gun, modeled after big six-shooters of the West. We bought and fired so many paper caps

156

that we soon found ourselves bankrupt. Next, he found some dry, straight-grained hickory and fashioned bows and arrows for us. He labored manfully to teach us archery, but we never acquired much real skill. Undiscouraged, he made us a crossbow, but it never worked well. It was something of an experimental model, and he never got around to correcting its faults.

Farm people found a good many uses for hickory wood in addition to making bows and arrows and smoking meat. They chose shellbark in preference to pig hickory. This hickory, when straight-grained and well seasoned, is quite hard and strong. It was cut into suitable lengths, then split and whittled to the approximate desired shape. Final form and smooth finish were produced by patient scraping with pieces of broken glass. Among items thus made were ramrods for muzzle-loading guns and handles for axes, mauls, mattocks, adzes, and the like. Hickory husking-pegs with a leather loop for attachment to two fingers of the right hand were widely used before factory-made steel pegs supplanted them. I once saw a much used tobacco pipe that a local smoker had whittled from hickory.

The St. Marys River flowed some twelve miles south of our farm. It was a small, sluggish stream, overgrown with brush and brambles, and its water was polluted by wastes from nearby oil fields. Although by no means a promising stream for the angler, we could not be satisfied until we got to go and try our luck there. The first visit, when our party caught nothing but a small carp or two, should have discouraged us, but it didn't. For a long time we looked forward in eager anticipation to the next trip. We even made a small wooden kit and put into it, with the fishing tackle, knives, forks, spoons, salt, and

157

pepper, so that we could cook and eat our fish fresh from the water. We took it along when we next went to the river, but it proved to be only superfluous baggage—no one caught a fish.

A sudden heavy rain one July day brought our creek brimming full and flooded low-lying adjacent ground. Like ducks, we made straight for those pools and started wading. In a low spot I felt something under my bare heel that I was sure must be a fish. Cautiously slipping my heel aside, I seized the "fish" between thumb and finger. Lifting it out, I found I was holding the head of a snake. That reptile must have been both astonished and immeasurably gratified at the suddenness and utter wholeheartedness of its release.

The best swimming spot we could boast was in a creek several miles distant. Although this was little more than a mudhole, it attracted groups of youthful swimmers every warm day. Big rains would quickly bring this creek bankfull of roily water. It was deep then, with a swift, powerful current, an exciting challenge to the best swimmers. A boy might learn to swim by watching his friends in the water or by listening to their talk between swims. Some, like pups or ducklings, just jumped into the water and swam. However the knowledge may have been acquired, practically every boy knew how to swim in some fashion. All went in naked—for years none of us so much as saw a bathing suit.

A doctor living in a neighboring river town had two small sons. Knowing that it would be impossible to keep them out of the water, he decided that for their safety and the parents' peace of mind, both must be taught to swim. Accordingly, he supplied them with bathing suits and went with them to the river, prepared to give them

158

a comprehensive lesson that thereafter would assure their safety in the water. When they neared the bank the boys— the elder only six—dashed ahead and, to his utter amazement, dived in and swam like fish.

One summer we talked Dad into buying us a bicycle. This was supposed to be a partnership machine, but my brother derived very little pleasure from it; he became disgusted with tire troubles, which began soon after we started using it. Its single-tube tires were attached to the rims by cement that would let go without warning. This allowed the tire to creep on the rim until the valve stem tore, setting up a leak that could not be repaired. More often than not, this happened when one was miles from home. A good many boys, singly and in little groups, rode for the fun of it; but bicycling was by no means an unalloyed pleasure. Not only were the tires subject to frequent mishaps, but the roads we had at the time were poorly adapted to riding.

As our house was centrally located, it was often the gathering place for a group of neighborhood boys on Sunday afternoons. A favorite pastime was hide-and-seek in and about the barn. Even more fun was pitching horseshoes. We put in many hours at this, and most of us became proficient at tossing ringers. Occasionally, we engaged in a Spartan contest with buggy whips: Two boys would stand facing each other, about two paces apart, each holding a long, lithe whip with which he lashed the other about the legs. The blows were laid on with increasing force until one or the other—usually the boy with the thinnest pants—called "enough."

In case someone knew where to find a nest of bumblebees, everything else was called off for the exciting pleasure of a "bee fight." These bees commonly established themselves in nests that had been built by field mice. No doubt, in most cases the mice had abandoned their homes voluntarily; in others, we suspected, they had been driven out by the bees. The nest might be in a clump of grass, under a pile of boards, in an old stump, or in a log. Usually we discovered nests by accidentally stumbling upon them or by kicking into something close enough to jar them. The colonies consisted ordinarily of a queen and a dozen or so workers. In the nests we found small leather-like pouches or cells, most of which contained larvae. In some were stored a few drops of crystal-clear honey; others contained beebread. We should have allowed the bees to go about their business unmolested, for they play a very important part in the pollination of clover and other useful plants. Ignorant and unmindful of this, we killed scores of them purely for the fun and excitement of fighting them.

Having located a nest, we armed ourselves with paddles and challenged the bees by punching a stick into the nest. The reaction was immediate, determined, and incredibly fierce. The bees, boiling with so much anger that one could smell very distinctly the characteristic bumblebee odor, poured forth to defend themselves and their home. They flew straight toward us at top speed, and we went into action with our paddles. When one saw a bee coming at him, he tried to swat it down; if he failed to see it coming, that was his hard luck. We learned that once a bee was down it was smart to crush it where it lay, provided that no other bee was attacking, for more than one of our warriors got stung by bees supposedly dead. In case a bee was not at once disposed of, it continued the attack until either it jabbed home its stinger or was put out of action. Occasionally, when there was no immediate de-

160

cisive gain by either contestant, the boy would take to his heels, slapping and batting furiously at the bee, in hot pursuit. It was at this stage, generally, that the bee scored, and the beefighter became a casualty. When the last of the insects had been disposed of, we would cautiously pull the nest apart, squeeze the few sweet drops from the honey cells, and swallow them.

Sooner or later each of us got stung. There was a sharp pain at the instant the venom was injected and for a few seconds afterward. For most of us there was no aftereffect except a slight temporary swelling at the punctured spot. One fighter among us was so susceptible, however, that a sting on his lip one time caused his entire face to remain grotesquely swollen for a full day.

I was once helping a neighbor in the hayfield when the loader carried up with the hay a nest of bumblebees. The infuriated insects flew at the men on the load like jet fighter planes, all meaning business. One fellow slapped wildly right and left with his hat; he was stung several times before he could roll to the ground. His brother stood calmly, not moving a muscle. The bees buzzed menacingly about him for several minutes, but he was not stung. Since then, I have followed his example, standing motionless upon encountering either bees or wasps; they never attack me.

As so many boys do, our little group fell into the error of believing that we had to learn to smoke. We tried rattan "cigars," cut from the stalks of old buggy whips. The slight appeal they had to begin with quickly ebbed after a puff or two. We gave corn silk a whirl, in the form of cigarettes wrapped in newspaper. Neither the taste nor the aroma satisfied, but we thought those fags did produce dandy smoke clouds. Dried leaves of catnip or pennyroyal weren't too bad in a pipe or cigarette. They had an agreeable smell and made plenty of smoke; but any thinking man would have agreed with us that the taste was anything but enjoyable.

The payoff came when we filled our clay pipes with some coarse blackleaf tobacco that we found curing in a shed at an old tile mill. This had just about everything except pleasing aroma, mildness, and agreeable taste. Some of the boys, when they recovered enough to think of tobacco without becoming sick again, started buying "Bull Durham" or "Duke's Mixture" and rolling their own. Now and then, someone would buy a pack of factory-made cigarettes, but his conscience pricked him cruelly at every puff. "Boughten" cigarettes, it was widely proclaimed, were "coffin nails"; and all of us had been warned repeatedly that they were among the most effective tools in the devil's comprehensive assortment. Some of the boys compromised by learning to chew tobacco—a majority of men about us were chewers; only a small number were regular smokers. Generally, they preferred pipes. Cigars as a rule were reserved for special occasions and were often used then by men who were not regular smokers. A good cigar cost a nickel. Some, offered at two for a nickel, were commonly called "two-fers." A few old ladies smoked pipes. Half a dozen boys became regular pipe smokers; several settled for hand-rolled cigarettes. A few tried earnestly, but they could not tolerate tobacco in any form. To mask telltale breath odor, some youthful smokers considered it politic to chew gum at times; others bought and used "Sen-Sen Breath Perfume" in the form of little black squares, sold widely in small envelopes.

162

FIFTEEN

Prevailing Fashions

FARM BOYS ALWAYS HAD A "Sunday" pair of shoes and a "Sunday" suit of clothes for relatively formal dress-up wear, principally on Sundays. When chill fall days brought reminders that winter was about to move in, my brother and I would be outfitted with cowhide boots for school and everyday wear. They had sheathlike, snug-fitting "legs," without laces or other fastenings, that rose almost to the knee. (It may be noted that from boots of this type were derived the terms "bootleg" and "bootlegger," used in the Volstead era in reference to the smuggling of liquor.) The thick, heavy soles and heels were thickly studded with steel or brass nails. Sometimes, instead of metal nails, they had wooden pegs; but, sooner or later, those with pegs got their full quota of nails. A few times we got boots adorned by red leather tops. Some boots had copper-armored toes; this was by no means a bad idea. Men commonly wore heavy leather boots for outdoor winter work. A few wore fine calf boots for dress at all seasons, the tops, almost knee-high, inside their trousers.

When new, our boots were black and shiny, the leather smooth and pliable. But soon the shine disappeared and wrinkles began to form, impressing themselves deeply and permanently in the leather from a point an inch or so below the ankle to a point several inches above. Each boot would accumulate three or four rings of wrinkles, making the ankle section look something like a concertina. Two straps at the top of each facilitated pulling them on. One could usually get new boots off fairly easily, but after they became wet a time or two and the wrinkles got a permanent set, someone had to pull them from your extended leg or you had to use a bootjack.

During all of our boot-wearing years, we had a bootjack in the house, within convenient reach. This device was strictly utilitarian, made from a piece of board about fifteen inches long. It had a V-shaped notch at one end, and a cleat was nailed to the underside, just back of the notch. To use it, you stepped upon it with one foot, thrust the counter of the boot on the other foot into the notch, and then pulled vigorously.

Without actually meaning to do so, we got the leather of our boots pretty well saturated with water just about every day. This commonly resulted from wading in water or wet snow just for the fun of wading. Occasionally, it came from accidentally stepping into a deep water-filled hole or from falling into some ditch or puddle and getting one or both boots filled with water. On taking them off at night, we would range our boots about the stove to dry. Sometimes they did dry a little. Invariably, though, when we came to put them on in the morning we found them almost as stiff and unyielding as if they had been made of wood. The drier they got, the stiffer the leather became. If the fire had not been going long and well, they were sure to be uncomfortably cold and clammy. From time

166

to time we greased the leather well with warm tallow or a mixture of lard and tallow, to which might be added a little pine tar. This treatment improved the appearance of the boots. It also restored some pliability and for a short time made the leather impervious to water.

Our boots stood up against an incredible amount of hard service. Getting them really clean, though, was extremely difficult. Furthermore, to use an expression long current locally, they were ugly as a mud fence. If these shortcomings were ever apparent to us—I don't recall that they were—our attitude toward the footgear was not affected at all adversely. We liked boots and they suited us to a T for the simple reason that nearly all the boys we knew wore boots just like ours. A boy's boots were supposed to withstand the wear and tear of a whole winter. They might do that unless he inadvertently got them too close to a hot stove and burned the leather, so that it cracked open, or unless they had to be resoled so often that they would no longer hold nails. In such a case he might be allowed to take his Sunday shoes for everyday wear; or he might get a new pair of "plow" shoes, which were hopefully intended to carry him through the remainder of the winter and, barring bad luck, well into or through the summer.

Plow shoes, supposedly, were to be worn when one plowed—if his feet were shod at all. They were designed to withstand hard knocks in any kind of work or play, being made of tough unlined leather with heavy, closely nailed soles and heels. They differed from boots only in that they came with buckles or laces and their tops were only ankle-high. For a long time few of the farm implements we used had any provision for riding. That meant a lot of walking. When the ground was wet our shoes became weighted with mud. Dragging that heavy footwear about gave a fellow considerably more leg exercise than he would ordinarily regard as essential. Working in loose, dry soil, more or less of it was scooped up and thrown into our big brogans. That, of course, was conducive to dirty feet and a long succession of dirty socks. Endless scuffing and kicking, attempts (favorable and unfavorable) to slide at every opportunity, and walking to and from school over our piked roads took an appalling toll of sole leather. In desperation a number of men in the community bought do-it-yourself cobbler outfits with generous supplies of nails and tough leather, determined, come what might, to keep their offspring shod. In general they were successful. Their handicraft was rough and very patently that of amateurs, but they managed to nail on soles and heels so firmly that they would stay nailed on. It may be stated for the record (if necessary) that those nail-studded boots and shoes, whether home-cobbled or cobbled by professionals, made frightful inroads upon floors and carpets.

For years oxfords or low shoes were worn only by women and girls, mainly in summer—they were far too sissyish for males. Dress shoes for boys were commonly fastened with buttons. "Congress" shoes or "gaiters," with elastic side gores and with prominent pull-on loops fore and aft, at one time enjoyed much popularity among men and boys. They were handy when it came to putting them on or taking them off—a moderate yank in one direction or the other sufficed. But, once the gore rubber failed, the shoes tended to give one a sloppy unkempt appearance. In addition to that, one or more of the pull-on loops would nearly always be found holding up, at a rakish angle, a portion of the bottom hem of the wearer's pants.

For a long time only black shoes and boots were worn. Only a few ever thought of polishing their footwear. These Beau Brummells used a polish that came in a tin box and was applied with a handled implement having a dauber on one side and a polishing brush on the other. The user first spat on the dauber or on the polish, then applied a coating of the waxy compound to his shoes. He finally went over them vigorously with the brush. This could produce quite a lustrous shine, but it tended to be a short-lived one—mud or water readily dissolved the polish. In dry weather the shoes quickly collected and stubbornly held a thick layer of dust. There was a real need for rubber overshoes for protection against mud and snow. Style-conscious young men frowned upon them—they made big feet look bigger and clumsier. Apparently, no argument against wearing rubbers carried any weight with me in my preschool years; they had an irresistible appeal after I had seen others wearing them. At the first opportunity I went with Dad to town and headed straight for a shoe store. There a pair of the coveted rubbers was fitted to my shoes. Handing the clerk my purse containing my meager accumulation of cash, I asked him to take out the amount due. It happened that the man was one of Dad's friends, and he thought the story too good to keep; so I was embarrassed many times thereafter by hearing details of the transaction recounted.

Most of the older men wore ankle-high, cloth-covered arctics over their shoes or calf boots. Men who did much outdoor work in winter often wore felt boots. Occasionally, boys were outfitted with them too, but most boys preferred leather-soled footwear because it was better adapted to skating and sliding. The felt boot was an extraordinarily clumsy affair, made of wool felt nearly half an inch thick and molded to conform roughly to the shape of the foot and lower leg. Canvas reinforced rubbers with buckle fastenings were worn over them. Though heavy and ugly, they were warm and comfortable, even in the coldest weather. But many complained that sooner or later they caused their feet to sweat excessively.

Some farm boys suffered from chilblains during part of every winter. This should have surprised no one because, until brisk exercise gets the blood into lively circulation, feet shod as ours were can become extremely cold when the thermometer stands well below the freezing point. In chilblains the tissues are not quite frozen; the after-effects, therefore, are not quite so serious as when actual freezing occurs. Nevertheless, I can state with the voice of experience that chilblains can be very bothersome. The trouble usually continued until well into spring, the affected areas itching, burning, and aching during prolonged periods each day. The discomfort was intensified if the victim forgot and held his feet too near a hot stove.

Toward the first of May our boots or shoes were likely to be completely done in. This never worried anyone much; May 1 was traditionally the time to start going barefoot. That was one thing that was done on the farm strictly by the calendar—boys stoutly insisted on that point. No matter how warm and pleasant April weather might be, no one thought of discarding his footgear—if he did, some higher authority was sure to veto the idea. But let May 1 come; then, even though the day might be cold and blustery, one could find barefoot kids cavorting everywhere.

170

It was a satisfying change, after a long winter, to go about our daily activities on unshod feet; but there were disadvantages: thorns, broken glass, nails, and splinters seemed to lurk in the most unexpected places, ready to puncture unprotected soles and heels. Moreover, the foot-washing routine was an inflexible daily pre-bedtime requirement. Throughout the summer, a wooden pail—we called it a "foot bucket"—was kept just outside the kitchen door for our use. For those nightly ablutions we used water pumped directly from the well. (For baths we depended, like nearly everyone else in those times, on water in a basin or tub and a sponge or cloth. The water was heated in pails on the kitchen stove or dipped from the stove's reservoir.)

One summer, after going shoeless for weeks, I developed what was called a stone bruise on my heel. It was a big black spot that became very puffy and, finally, extremely sore. Considering then that it had become "ripe," Dad lanced it with his razor, cleaned it well, and applied a bandage. That brought marvelous relief, and within a short time it healed completely. Once I accidentally dropped a heavy file. Its tang, like a very blunt spear, pierced the unprotected nail of my big toe and went on through the appendage. For a long time I was not only a barefoot boy with cheek of tan but I could also have been identified by an incredibly sore toe.

In our earlier years harvesttime sometimes found us with nothing to wear for the protection of our feet against the sharp stubbles of the hay or wheat field or from dry thistles that might be encountered. The soles of our feet by that time were so thickly calloused that they had a leatherlike toughness, but the sharp stubbles scratched the skin of ankles and legs unmercifully. And how those scratches did

171

smart and sting when we washed our feet before retiring at night!

One year buttoned, cloth-topped shoes were the height of fashion for males. Some time later, after oxfords had come into general use by young men (for summer wear only), all who made any pretense of being stylish dressers wore them with red laces. This was about the time that all young fellows who really knew what was what in fashion were wearing derby hats (preferably brown), gaudy vests, black coats, light gray pants, and extremely high stiff collars that a sensible giraffe would almost have balked at.

Nearly all of the clothing worn by small boys was homemade. Often, their pants were made from cloth cut from the soundest parts of garments discarded by their dads. Men bought their work pants and overalls ready made. Socks were factory-knitted but men's work shirts—even their underwear in some instances—were homemade. The first "store" pants we boys had were made from a tough cotton fabric with triple-sewed seams and riveted buttons. Accompanying them was a folder that glowingly set forth their super-excellent quality and announced that the dealer would "cheerfully" give the buyer a new pair if they ripped, and pay him a quarter for each button that came off. One day, after we had worn the garments almost to tatters, Dad was setting off for town when my brother came running with his worn-out pants wrapped in a newspaper. "You know they were guaranteed," he said. "Take 'em back and make the man give me a new pair." Feeling that the dealer would not do that very cheerfully, Dad managed to talk him out of the idea.

Attached collars for dress shirts were practically unknown. All such shirts had neckbands made for attaching stiff linen collars by means of inserted collar buttons. All were white, generally with broad, stiff bosoms with tabs for anchorage to a button on the trousers waistband. Here and there, a few celluloid collars and cuffs appeared; but all were so shiny and so patently imitations of the real thing that they never gained widespread popularity. One could buy fancy, ready-knotted "teck" ties. They offered the undeniable advantage of being easy to put on, but many considered the knots too perfect. Besides, they had a droll way of coming loose at inopportune moments and lopping down, wrong side out, over the wearer's front.

One winter, "Way" mufflers appeared in all the clothing stores. They caught the fancy of men and boys and not a few girls. The best were made of fine knitted wool with a thick, warm fold to encircle the neck over the regular collar and fasten at the back with snap buttons. A wide "bib" in front tucked inside coat or vest. All were designed to make it appear from the front that the wearer had on a turtle-neck sweater inside his coat. Few were interested in this make-believe quality; the muffler's popularity was due mainly to the warm comfort it afforded and to the convenience of putting it on or taking it off.

At that time women and girls were wearing "fascinators," which had been in use long before Way mufflers appeared, and continued in favor long after the latter were forgotten. The fascinator was usually made of loosely crocheted wool yarn, in the form of an isoceles triangle with a broad base. It was worn, folded a time or two, over the head, the ends wrapped about the neck.

Almost everybody wore heavy knitted underwear in winter, nearly always two-piece suits, with full-length sleeves and legs. Men who worked outdoors in winter often wore two pairs of pants or a pair of rugged overalls over heavy pants. Over vests and blouses they wore roomy denim jackets, often blanketlined, with an interlining of oilcloth to shed water and break the wind. For everyday wear in winter men and boys bought heavy mittens made of ticking, lined with a fleecy material and having a cotton-felt layer between ticking and lining.

At one time a few men wore round, black felt hats with flat-topped crowns. They were not railroad men, but the hats were known as railroad hats because some engineers and other railroaders wore them at their work. In winter farmers and farm boys favored a sort of lumberman's cap for work and dress wear. This was made of heavy woolen cloth with a moderately long peak and folding ear flaps.

Suntan had not come to be regarded as an essential attribute of feminine pulchritude or of rugged manhood. Women and girls went about their outdoor duties on the farm in summer wearing big, homemade cloth sunbonnets that sheltered faces completely from direct solar rays. Men and boys had broad-brimmed hats of coarse straw for wear as they went about their everyday affairs. In spite of this protection the faces, as well as hands, of older men, exposed to sunlight during long periods every day, often took on a mahogany tan by midsummer. A few young men wore canvas or goatskin gloves at their work throughout the summer. This was to keep hands soft and white and thus, supposedly, put their owners in an elite class.

Boys generally had to be satisfied with "homemade" haircuts. A number of boys and young men patronized a self-taught barber in our neighborhood. This fellow actually cut his own hair (it was done with mirrors, to a considerable extent). Something like a third of the older men

174

wore mustaches. Some kept them neatly trimmed, but the majority allowed them free rein, and they tended to assume the walrus or handlebar form. Often in winter icicles would form on long mustachios, giving wearers an absurd, grotesque appearance.

A good many men of fifty and older wore full beards; a few shaved the sides of their faces regularly and grew only chin beards and mustaches. Often the effect produced when bewhiskered fellows chewed tobacco fell far short of being an aesthetic asset. And how a full-panoplied mustache or a grizzly-like growth of beard did alter a man's appearance! Conversely, shaving off a growth that had been cultivated a long time might produce an effect even more striking.

SIXTEEN

Callers at Farmers' Doors

EARLY EACH SPRING THE "assessor" with his flat tin box filled with tax papers appeared at the farm. He was an elected township official, usually a well-known neighbor. He and Dad listed together grain, hay, livestock, and other property that was subject to taxation. Together they arrived at appraisals on which county officials would later levy the tax. Since both were disposed to be fair and honest, there was no disagreement, no haggling.

Occasionally, a nursery salesman appeared soon after the assessor. He also carried a flat tin box. It contained catalogs and circulars with gorgeous pictures in color of amazingly large and perfect peaches, apples, pears, plums, etc. The man would talk expansively of the high quality of his stock and point out just where certain choice varieties should be planted. He would dwell at length upon the landscaping improvements that would accrue from having them there, and picture in glowing terms the enjoyment that would come to the entire family when the trees and vines came into bearing. His customers heard him with respect, because he was regarded as an authority on fruit growing. He did some grafting occasionally, using a grafting wax that he made by combining wax produced by his own bees with tallow and linseed oil. He carried a huge pruning knife and, at times, as he talked, trimmed water sprouts from our big apple trees. Now and then, he would seize an ear of one of the listening youngsters and pretend playfully to cut it off. This was rather exciting for we were never altogether certain that he was jesting.

Salesmen with lightning rods appeared often. Some were clearly swindlers, with tricky contracts and decidedly inferior products. Actually, some of the rods, due to improper construction and installation, were worse than none. Peddlers with heavy packs suspended by shoulder straps came trudging in, half a dozen or more of them every summer. People generally called them "Arabs." They had a strange, foreign appearance; all were dark-skinned with black beards, and all had difficulty with English. Each carried a varied assortment of trinkets and notions, all cheap and tawdry. Occasionally, one would offer to exchange a few baubles for a night's lodging; but in all cases these offers were rejected—most people were suspicious of them. Generally, it was believed, they crept into haymows or stacks and bedded down there.

Occasionally, a peddler would come with his wares in a spring wagon drawn by a decrepit horse. In addition to notions and cheap jewelry he carried scissors, harmonicas, spectacles, shawls, and an assortment of tinware. He tried to sell for cash but, failing that, would trade for bags of rags, metal scraps, dry bones, or other junk material.

Three or four book salesmen—all "agents" to local people—worked through the community every year. The real salesmen among them must have found it a lucrative business. Their books were generally of the subscription type,

178

planned and produced to be sold by door-to-door salesmen, who carried only a glowing "prospectus" and took orders for later delivery. Some were religious works; others were represented as educational books, crammed with the latest important, essential information and therefore of the greatest value to both adults and children in school.

Many of those books were quite mediocre and often of questionable authenticity because they had been compiled from obsolete sources by writers who knew little of the subject treated and apparently cared less. A few were really good, yet any intelligent person could have selected much better ones from the catalogues of standard publishers and saved money by so doing.

At that time books as sources of information and entertainment faced nothing like the competition that came later from magazines, the daily press, the radio, movies, and finally television. A goodly number of people were aware of their own educational deficiencies and had a real thirst for knowledge. The appeal to parents who were deeply interested in having their children progress educationally was well-nigh irresistible. Because they didn't know how to choose and buy books through regular channels, and because for a long time no public library was available, many farm people welcomed the salesmen who brought books to their doors.

One year, salesmen covered the area and talked almost every family into buying a stereoscope with an assortment of views. The instrument and the pictures for a long time occupied a respected place on parlor tables. Traveling photographers with equipment in darkrooms built into their wagons persuaded many farmers to have their buildings, livestock, and families photographed. They visited the district schools, got teachers to line up the pupils, and turned a few honest dollars by photographing the groups. Occasionally, salesmen covered the territory taking orders for life-size enlargements of album photographs. The most promising prospects were relatives of recently deceased persons. The pictures they delivered, though rather crude crayon portraits, were generally accepted as satisfactory.

One day a salesman demonstrated to us a combination device that could be used for stretching wire, erecting fences, or hoisting. Convinced that the tool—well designed, compact, and handy—would be useful, we ordered one. It turned out to be totally worthless because the manufacturer had used cast iron instead of cast steel for vital parts. We found that out when one of the castings snapped like glass under a load of only a few pounds. Since it would have cost only a few cents more to make the parts of steel, we figured that either the maker was far too intent on low-cost production or he didn't know any better.

SEVENTEEN

The People of Black Swamp

A FEW COMMUNITIES IN THE
region were originally settled mainly by people of Welsh
descent. Here, both English and Welsh have long been
used in ordinary communication. Welshmen are lovers of
music, and many have special ability as singers. The use
of the Welsh language is now gradually diminishing, but
singing, public and private, is given widespread encouragement and remains a primary interest. Observers generally
would agree that this is about the only characteristic that
marks the Welsh as different from neighbors of other
racial extractions.

Other settlements began almost as miniature German
provinces. During a long period, few spoke English.
Church services were conducted in German, and children
in school were commonly taught in the mother tongue.
Succeeding generations, due to intermarriage and infiltra-
tion from outside areas, gradually abandoned speech and
habits that marked them as Germans. Little now remains
except family names to hint of German origins. Many of
the first settlers in a few other communities were of Swiss
origin. Qualities that from first to last have most noticeably differentiated them are exceptional industry, thrift,
and competence at farming and other pursuits.

Among the people about us—on the whole, I believe, typical of the inhabitants of the Black Swamp region—were representatives of English, German, Irish, Scotch, French, and Welsh stocks; but one rarely heard a word of any language but English. Practically all were several generations removed from the Old World. A large proportion were descendants of early settlers of the area who had come from older Ohio counties. Many of the latter in turn were scions of pioneer families of other, older American states. Some could have traced their ancestry back to Europeans who were among the first to establish themselves in the New World.

Nowhere, probably, could one have found a more striking example of the successful functioning of the American melting pot. No Old World nationalism, prejudice, or ideology was in evidence; all were good American citizens and friendly, co-operative neighbors. Practically all were intelligent, honest, industrious, and thrifty. The majority were churchgoers. Among people we knew, Republicans outnumbered Democrats, but that was taken as of little local significance. Almost to a man they could have been classified as rugged individualists. They stood on their own feet, and they wanted it that way. In trouble they stood together. When sickness came, or disaster, help for the unfortunate family or individual was always forthcoming, help prompted by sincere and sympathetic hearts that neither counted the cost nor looked for recompense.

Much of what these people knew had been learned through hard knocks experienced by themselves and their fathers before them. While they had respect for tradition, old methods, and old customs, few revered the old for its age alone. On the other hand, few were disposed to

182

welcome the new unreservedly merely because it was new, although most were ready enough to abandon the old and accept the new when the latter had proved itself sound and worthy of acceptance.

Nearly all came from farmer ancestry, and their interests were centered about farms and farming. Land was their capital and their insurance against the inevitable day of retirement from productive labor. About all carried fire and tornado insurance, their policies generally being written by mutual companies of which they themselves were members. Few put any money into life policies.

Food materials were abundant; yet many regarded the deliberate destruction or waste of anything of this kind that could have value for any purpose as inexcusable—even sinful. Most people had a wholesome respect for money, even pennies. Very few, however, could justly have been called miserly or niggardly. Money was hard to come by; they had to get along on cash incomes that today would seem meager in the extreme. Many things now regarded as necessities were then luxuries that could be dispensed with, and many of today's luxuries were unknown.

Practically all had a deep-seated fear of debt, having been taught by precept and example that often "Poverty rides on Debt's back." Many, however, at one time or another faced debts, incurred in buying farms or needed equipment. Once in debt, they strove to whittle their indebtedness down as rapidly as possible. In the eyes of most people we knew, nothing so distinctly marked one a failure as arriving at old age penniless, due allowance of course being made for circumstances. Contemplating their own sunset years, they felt that nothing, when those years came, could be quite so dreadful, so humiliating, or

183

so shameful as to find themselves at last homeless, bankrupt, and dependent, going to the poorhouse for shelter
and a place to die. Judgments tended to be harsh is cases
where a fate of this kind clearly resulted from an individual's own laziness, foolishness, or improvidence. The
ultimate unhappy predicament of the indolent, "do-less"
fellow was likely to evoke more mirth than sympathy.
Many laughed at an oft-told tale of a shiftless old character
known as Tommy. Neighbor A and Neighbor B, in their
wagons, met on the road. As they exchanged greetings,
A, seeing old Tommy seated on the floor of B's wagon,
inquired:

"Where's Tommy going?"

"Says he's too old to work any more," replied B. "Not
much to eat at his place, either. Taking him to the poorhouse."

"Hey, got a bag of corn here," said A. "Tommy can
have it. Keep him going a while yet."

At this Tommy eased himself up lazily, peered over
the sideboard at A and asked, "Is that there corn shelled?"

"No, Tommy."

The old fellow turned to B with an air of resigned
finality and, as he settled himself as comfortably as he
could on the hard boards, drawled in a listless voice,
"Drive on, then."

Most farmers worked early and late to get their farms
well fenced and well drained, to keep their soil in good
tilth, and to provide comfortable homes, good barns, and
so on. They tried to lay by some money for the purpose
of aiding their children to get a start in the farming business when they married. This became increasingly difficult
as the population grew, as land values went higher, and

184

as share-the-wealth legislation became operative. Ordinarily, a young man with an ambition to own a farm started, with what financial help his parents could provide, by buying horses, livestock, and necessary farming equipment for the operation of a rented farm. If he managed well, he would be able within a few years to become a full-fledged land-owning farmer. In pre-mechanization days the average farmer was content if he had eighty to a hundred acres, about all that he could handle well.

Generally, a farmer's children counted a great deal more as assets than as liabilities, for they aided with innumerable tasks that had to be done. Boys were assigned outdoor jobs; girls assisted their mothers. In case there were more girls than boys, the girls helped with work indoors and out. If, on the other hand, boys predominated, inducing them to do housework was likely to be a difficult undertaking—often a hopeless one. In a few homes the children were allowed to grow up without much attention to standards of conduct or moral or ethical principles. In most other homes the parents were firm believers in the adage "Spare the rod and spoil the child." Actually, the rod was used only on rare occasions and as a last resort. In no observed instance did the maintenance of discipline in such homes give rise to permanent resentment or bitterness on the part of any child.

My father and mother and many of their contemporaries used the names "Pap" and "Mother" in speaking of, or to, their parents. A few used "Father" instead of "Pap." We children and others of our generation were taught to say "Pa" and "Ma." Occasionally "Papa" and "Mama," abbreviated to "Pop" and "Mom," were heard. Late in the 1890's, nearly all of the boys of our community, with surprising unanimity, began substituting "Dad" for "Pa." There was no known objection from fathers to the newly adopted name. Apparently, they sensed that it signified no disrespect or lack of affection and that the kids were just following the popular trend. At this time "Mom" enjoyed increased popularity. Girls showed some reluctance at first to adopt the innovation, but soon most of them fell into line. The change seemed to come suddenly and spontaneously—helped along no doubt by the herd instinct. From that time until this day, it appears, a considerable proportion of American fathers have been "Dad" to their offspring and mothers have been "Mom." Exactly how this custom originated and what caused its rapid and widespread adoption have always puzzled me.

Until the children grew big enough to help, most farmers had to hire hands, particularly in the busy seasons. Some employed men the year around. A few families kept hired girls throughout the year, but in most cases jobs were open for girls only in the busiest seasons or when a new baby came to impose additional tasks in a farm household. In no case did employment as a farm hand or as a helper in the house appear to lower caste or reflect upon these young men and women. They were treated as members of the family, not as menials. A young man who came one time to service our parlor organ reflected something of the prevailing democratic attitude in this connection. He told of having "kept company" with a young lady who entertained what he considered high-toned, aristocratic ideas. "When I marry," she told him, "I'm going to keep a hired girl. No drudging for me."

"When I marry," came back the young man, "I'm going to marry the hired girl."

No farmer was very rich, few, very poor. Among the wealthiest were men who had inherited good farms and managed well. Oil from wells on the property of some put them on Easy Street. Although most of the real wealth was produced on farms, a considerable part of it had a way of gravitating to town. Owners of grain elevators, most of whom had moved to town from farms, as a rule did very well indeed. They bought farmers' products, then sold them needed supplies, making profits when they bought and when they sold. Hardware stores, lumber yards, and implement dealers enjoyed a lucrative business with farmers, Bankers, most of whom contributed constructively to the development of the area, fared well in dealings with farmer clients.

The fortune of one of the town's richest men came primarily from the timber that farmers hauled in and sold to him. As owner of a factory that made wood products, he was among the largest employers of labor in the entire region. He contributed liberally to local charities, but this did not serve to allay the general feeling that he was a heartless driver of the men who worked for him, exploiting them and robbing them of rightful fruits of their labors to enrich himself.

A man in town, who fitted almost perfectly the description of Longfellow's hero who labored in a smithy under a spreading chestnut tree, shod our horses and did all our blacksmithing. He was a fast workman, putting in long hours every day, but jobs flowed in so fast he could never get ahead of them. Working altogether with hand tools, he could make and repair practically anything in wood or steel. He could build a sound, sturdy wagon, but he rarely got a chance to do so; farmers kept him too busy

187

setting tires, making new wheels, tongues, spokes, felloes, hounds, and reaches for their old wagons. Jointers and shares of plows had to be sharpened from time to time. Points wore down so that new ones, commonly made from worn-out, hard-tempered rasps, had to be welded on and forged into shape. A large amount of such plow work fell to the lot of our blacksmith.

Another local blacksmith built a few wagons that farmers liked so well that he had to give up about all of his other work and specialize in wagon-building. In time he began building buggies also, and eventually converted his shop into a small manufacturing establishment. Wagons built by Studebaker, later a manufacturer of automobiles, were widely used in the area. Others came from other big factories, some in Indiana, some in Ohio.

No one appears to have been very consciously influenced by the couplet "Early to bed and early to rise, makes a man healthy, wealthy, and wise." Yet early rising was an established custom. When one began early in the morning and put in a long, full day at work, it followed that he had to retire early in the evening to get a full quota of rest and sleep. The average bedtime was around nine P.M. Many arose regularly at five each morning and started the day's activities, much of the time by lantern light. If a man failed many times to get out at about seven in the morning to work with his team in the field, after doing the milking and other chores at the barn, his neighbors were likely to wonder whether, after all, he might not be a little lazy and "do-less."

Waking lazy farm hands and sleepy youngsters to get them out in time for breakfast was often something of a

188

problem. One man had a hand who seemed to be lulled into sounder slumber at each successive morning summons to rise. One morning, after this had gone on for more than a week, he called repeatedly without results. Exasperated, he went to the young man's room, threw down the covers of his bed, and dashed a cup of cold water over him. This had an electrifying effect. It made him a little angry, but he managed thereafter to get up at the first call. Another man had two strapping sons who were not unduly fond of work but had a passionate love for hunting. Rumor had it that their dad sometimes resorted to a ruse to get them stirring. "Boys," he would call, "get up—fine mornin' for quail."

Probably it could not be proved that early retiring and early rising actually contributed greatly to the health, wealth, or wisdom of anyone thereabouts. There is no doubt, however, that under conditions that made so much human labor necessary, those who started early in the morning and worked industriously through long days prospered as they could not have done otherwise.

We never knew of a Christmas tree being set up in any home in the community. There were family gatherings, though, and always bounteous dinners in the true Christmas tradition. Gifts were provided for all, especially the children, but they did not appear in anything like the profusion of later times. They tended to be simple and comparatively inexpensive. Some practical-minded parents inclined strongly toward giving useful, needed items, such as articles of clothing and school supplies. Youthful recipients of such gifts naturally took a dim view of this, feeling that sooner or later all would have come to them,

189

Christmas or no Christmas. We children found our gifts on Christmas morning in shoes or boots or in stockings nearby—no one we knew hung up his stockings. Fancy gift wrapping was unheard of. Holly, mistletoe and special decorations rarely appeared. No one we knew had a yule log for few fireplaces were then in use.

No one in our family ever had a birthday party or a birthday cake, as such. Birthdays were observed only to the extent that each of us could always count on a birthday "licking" from brother, sister, or schoolmate, one resounding whack for each year, plus "one to live on and one to grow on." Birthdays were accorded similar treatment in other homes except that occasionally one of our friends, on his own initiative, would get up a party for himself.

Church weddings were somewhat rare; most couples were married in the home of the bride. The bride might have a bouquet, but, as a rule, no other flowers appeared. Guests wore their best Sunday clothes—no one had formal attire. One time when I was about twelve, I was helping Uncle Cal gather and burn fragments of stumps in a newly-cleared field. On the eve of a wedding at the neighborhood church, he invited me to go with the family. I demurred, arguing that I had nothing to wear except my rough work clothes. Since everyone else wanted to go, I found myself talked down and dragooned into going.

Uncle Cal seated me beside him in a prominent place near the front of the auditorium, my coarse shoes, dusty pants, and work-stained jacket standing out conspicuously. In the midst of the ceremony, I thrust my feet under the seat and knocked over the janitor's dustpan and broom

190

that, by some devilish chance, had been left there. The clatter was horrendously loud and distracting, but no one seemed to mind it half as much as I did. At that moment nothing could have pleased me more than for the floor beneath me to suddenly collapse and drop me into the dark, neglected basement, where I might lie buried under debris, the deeper the better.

Nearly always, following a wedding, refreshments were served at the home of the bride; or there might be an elaborate dinner. In some cases parents of the bridegroom, some time later, would honor the couple at a reception marked by a gorgeous dinner at the young man's home. Such receptions were known as "infares."

No wedding was complete without a belling or "shivaree." At a home wedding, the belling might come shortly after the ceremony. If the couple managed to give the serenaders the slip, they were pretty sure to be belled soon after they returned from the honeymoon. Cowbells, horns, tin pans, conch shells, horse fiddles, and shotguns—sometimes even dynamite—were used to make a frightful din. Occasionally, someone carried in a dinner bell or a buzz saw to be beaten with a heavy club. In most cases bellings were stag affairs, but at times girls joined in the fun. The serenade continued until the bridegroom appeared and passed out cigars, with candy or other confections if girls were in the crowd.

Most conch shells were family heirlooms that at one time had been used to announce dinner to men working in the fields. In our time only the dinner bell was used for this purpose, the shells being kept as curios or ornaments that could be pressed into service as door stops. The dinner bell, of cast metal, with a rope attached to its crank, was commonly mounted on a post near the kitchen. The post was supposed to be of such a height that the bell could be heard at far corners of the farm. The bell was used not only to summon men to meals but to sound an alarm in case of emergency. One could identify all the bells in the neighborhood by their tones. The need for them having diminished or ended altogether, these bells are now fading from the farm scene, going to serve as "door" bells for town houses or to become items in antique collections.

There were no funeral chapels such as those of later times. Practically all funeral services were held in churches, preceded by a solemn tolling of the bell. Nearly always, the music, provided by a small choral group, was as sadly funereal as could well be imagined. The funeral sermon was rarely less than an hour in length.

It was customary for a group of friends and neighbors of a bereaved family to sit up all night in the home where the body lay. At such gatherings, which were never referred to as wakes, the visitors entertained themselves by low-toned conversation, which at times might become quite lively. Nearly all napped a little now and then. Pie, cake, and coffee were served. There was never any liquor, never any boisterous or unseemly conduct.

Several men and youths of our community were widely known as swappers. They were never happier than when they could work up a deal with someone, and the more "boot" they could wangle, the happier the transaction made them. There was little actual misrepresentation or cheating, but such clever psychology was applied that

192

one who swapped with a real expert would do well always to keep his wits about him. Even with that precaution, he stood in danger of losing his shirt. Most of the trades were pure barter, in which one exchanged on a fair basis some article he no longer needed for some other that would be useful. If the items swapped were not clearly of approximately equal value, there might be a great deal of jockeying and parleying to reach an agreement on relative values and determine just what amount in boot one trader should receive from the other.

One of our neighbors, a man named Jesse, was quite fond of swapping. He was so good at it that several times professionals who swapped with him came out second best. He came one time offering Dad a big key-wind watch for a small stack of hay. It happened that Dad didn't need the hay and, as he then had no watch, they made an even trade. That heavy Waltham was an excellent timekeeper. Dad carried it daily for many years. A short time after this transaction, Jesse traded Dad a calf for a pig. A few days later, Jesse passed the field where Dad and his hired man, John Coon, were husking corn. Dad reported that John was calling the calf "Jesse."

"Pig's named, too, Bill," replied Jesse. "We call him Billy Coon."

At school one day I swapped a fountain pen for a watch. The thing was running all right, ticking like an alarm clock at the time the exchange was made. When I pulled it from my pocket a few minutes later, it had stopped dead as the most lifeless of mackerels. I found that if I gave the case a quick clockwise turn and held it horizontally, it would run for several minutes—that explained how I had been taken in. I contacted Jesse, and we soon arranged a swap, the watch for a .22 caliber rifle. When I later fired the gun, powder blew through a hole burned in the breechblock and scorched my face, making me gunshy for months. Both parties to that deal were cheated.

Not long after we moved into the new house, a friend of the family who sold musical instruments drove in with a new parlor organ. Whether by accident or by design, that visit was most happily timed so far as sales possibilities were concerned, for our parents had been giving serious thought to getting the three of us started with music. The salesman brought the instrument into the house, demonstrated its tone, and launched into his sales talk. When he named the price, Dad admitted that it seemed fair enough but countered with a proposal that a fodder chopper, which he no longer needed, be accepted in a trade. The man conceded that he could use the machine, since he lived on a farm. So the organ was swapped for the chopper, plus an agreed amount of boot.

194

EIGHTEEN

Language of Everyday Communication

ALTHOUGH ENGLISH GRAMMAR
had been taught for a good many years in rural public
schools, much of the language heard about us was far from
grammatical. There were some, including a few ex-teach-
ers, who could speak good English, but in everyday conversation they tended to fall back upon the homely lingo
of the majority; it seemed that they didn't want to be regarded as stuck-up. In our school—probably fairly typical
of regional rural schools—a sizable proportion of each class
approached grammar with almost total indifference. To
some extent, at least, this appears to have reflected parental
attitude—one boy among my classmates, encouraged by his
father, refused to have anything whatever to do with the
subject.

Probably few of the words and expressions in common,
informal use were actually indigenous to the Black Swamp.
For the most part they were imports, having been brought
in by settlers who came from outside areas, far and near,
or having been picked up from time to time by local people
in contacts with inhabitants of other regions. Naturally, a
fairly large number of colloquial words and phrases applied
specifically to the business of farming. A good many were

pointed and picturesque. Some figures of speech had a grotesque quality, some a quaint tinge of humor. References to proverbs and epigrammatic sayings that are the common heritage of users of English were frequent.

In local parlance anyone or anything considered quite insignificant or of little worth was "triflin'," didn't "amount to nothin'," or, perhaps, might be "the last joint on the tail end of nothin'." At the opposite extreme one "set store" by something highly prized or greatly admired. If there was something he disliked, he might try to "get shet" of it.

One who failed to get on well because of obvious laziness or other personal fault was a "no-account," or he didn't "amount to a hill of beans"; in an extreme case he might not "amount to powder and lead to blow him up." A man regarded as lazy was likely to be classified as "not very work-brickle" or as "do-less." Some would say of a person naturally slow and deliberate in action that he was "too slow to catch cold" or that it was necessary to "set stakes to see him move."

Of an individual considered very stubborn, some would say that he was "stubborn as a mule"; others might assert that "if he fell into a river he would float upstream." A closefisted, grasping man, it was said, was "tight as the bark on a tree" or "would skin a gnat for its hide."

When a man or boy wished to assure another that he was giving him full moral support, he said: "I'll stand your backin' till your belly caves in." The fellow who asked for quarter in any rough contest "hollered cavey." A boy, boasting of his prowess at fisticuffs, might assert that he could lick a prospective antagonist "with one hand tied behind my back." The classic response, in case the lad appeared lacking in physical ability to make good his boast,

196

was: "Yes? Where do you bury the ones you kill?" If someone, especially a boy, proposed for himself a task likely to
tax his ability, he might hear the remark, usually banter-
ing: "I'll bet; you'll do wonders and cat blunders." A playful threat of physical violence might be couched in some
such form as "I'll tan your hide," "I'll warm your jacket,"
or "I'll whale you".

Occasionally, someone said: "I like t'died." This could
mean that he had been greatly embarrassed, highly
amused, or ill. Now and then, we heard some moderately
well-to-do man say: "It's no disgrace to be poor but it's
kinda unhandy." One regarded as seriously lacking in
worldly goods was "poor as Job's turkey."

When one agreed to work for another, he "hired out." An
employer "turned off" a worker when he discharged or fired
him. To "hoe a crooked row" was to labor under disadvantages. Anyone who sought to speed up work by de-
manding extra effort from others was "cracking the whip."

Even as the boss in a city factory may send a green ap-
prentice for a "board-stretcher" or a can of "striped paint,"
the rural joker might assign a naïve helper to look for
"spotted pig tracks" or a "left-handed monkey wrench."

Questioning the soundness of an individual's judgment
and understanding, some averred that he "didn't know
beans" or "didn't know putty." Others, putting it more
bluntly, declared that he "didn't have sense enough to come
in out of the rain" or that he "didn't have sense enough to
pound sand in a rat hole". Referring to one who appeared
unobservant, some opined that he was "blind in one eye
and couldn't see with the other."

A girl's boy friend was her "feller" or her "beau". In case
she dismissed the chap, she "gave him the mitten." If, on

197

the other hand, they married, they "got spliced" and began "trotting in double harness." One was left to "dance in the hog-trough" if a younger sibling married before he did. A wife regarded as dictating family affairs "ruled the roost" or "wore the britches." A wife became a "widow woman" if her husband died.

Confronted by a problem that appeared extremely difficult, some said they were being "stumped" or "up a stump." A man who "figgered," "calculated," " 'lowed" he would do a certain thing was proposing or planning to do it. If he meant to waste no time about it, he said it would be done in "about three shakes of a dead lamb's tail." In case the proposed undertaking might appear a little pointless or silly, he said he would try it "just for devilment."

A man working very hard felt that he was "working like a dog." One subject to continued harassment and trouble was said to "live a dog's life." A notably dishonest person might be described as "crooked as a dog's hind leg" or referred to as a "slippery customer" who "could hide behind a corkscrew." Anyone who tackled a job vigorously "sailed into it"; if he stubbornly refused to give up in spite of all discouragement, he "held on like a pup to a root."

To "eat high on the hog" was to live well. "Hogging off" could mean either slipshod harvesting of a crop or allowing hogs to eat a crop on the ground where it grew. To go all out or commit oneself unreservedly was to "go the whole hog." One who put on a front or made an ostentatious display was "sticking a fat hog."

To talk irrelevantly or to argue from a false premise was to "bark up the wrong tree." When something altogether unexpected happened, total surprise was expressed by saying, "You could have knocked me over with a feather."

198

One who found himself in a very embarrassing situation said, "I could have crawled through a knothole." A chap who had been given a sharp reprimand said the fellow who gave him the treatment "gave me Hail Columbia" or "read my title clear." One who was held rigidly to strict rules said he had to "walk the chalk." When someone was thrown altogether upon his own resources, folks said it was a case of "root, hog, or die" or that he had to "paddle his own canoe."

A few times it happened that someone, feeling sure of the identity of a caller who had knocked at the door, responded with the inelegant flippancy occasionally used when members of the family or intimate friends were involved: "Come in if your nose is clean." Immediately afterward, he suffered acute embarrassment when the visitor proved to be a dignified stranger.

Meat at all tough was "tougher than a boiled owl." An article fresh from the store was "brand splinter new." Anything that seemed exactly right for a certain use was "just the kicks." If something upset a plan or made a project go awry it "played hob" with it. To "get on with the rat-killin' " was to proceed with the work at hand. In a heavy downpour it might be either "raining pitchforks with saw-log handles" or "raining cats and dogs."

A person who optimistically anticipated the successful outcome of an undertaking was "counting his chickens before they were hatched." Some, feeling that the game was not worth the candle, would declare: "It's too much sugar for a cent." Now and then, a fellow with a decided conviction might offer to "bet two cents and a brick watch." To indicate that any one of several procedures could be used to a given end, a more or less standard statement was:

"There's more than one way to skin a cat." Anything moving menacingly toward one was "making for him."

Anyone who became haughtily indignant was "riding a high horse." He "ripped and snorted" in case he lost his temper to the extent of resorting to violent speech or action. One who seemed to show his authority unduly was said to be "feeling his oats." The chap who appeared to have an exaggerated opinion of his own importance was "too big for his britches." The urbanite was a "dude" or a "town feller." One who lived in the country was a "sod-buster," a "country jake," or a "clodhopper." If either happened to be youthful and appeared forward or over-confident, he was likely to be labeled "brash" or "fresh."

In case a visitor came unexpectedly at meal time, he might be a welcome guest at the table, but with the distinct understanding that he had failed to "get his name in the pot."

A baby was "nussed" either when it was suckled or when someone held it in his arms. A crying child was "singin' broadmouth."

Housewives straightening a disordered room might say they were "redding it up." As applied to laundry work, rinsing occasionally became "renching." As many expressed it, teachers "learnt" children, they rarely taught them. To these people a child in school was nearly always a "scholar," not a pupil or a student. A few called an axle an "ex."

"Let alone" was extensively used as the equivalent of "not considering" or "not counting." Occasionally, we heard "ary" used for "either" or "any," and "nary" for "neither" or "not any." "Tit for tat" commonly served as the equivalent of "an eye for an eye." "Liable" and "apt" were used,

200

more or less interchangeably, to indicate likelihood or probability. A few habitually applied the adjective "immense" to about everything that to Theodore Roosevelt would have been "bully." "I reckon" was widely used in lieu of "I think," "I believe," "I presume," or "I assume"; "I reckon so" in response to a question was equivalent to "yes." "Expect" and "suppose" were generally overworked, being substituted for "assume," "presume," "believe," or "imagine."

Words ending in "ow" were often given instead an "er" termination; they became "beller," "holler," "winder," etc. Numerous words were misspelled and mispronounced. Some of the most common were rhubarb (ruberb); radish (reddish); jaundice (janders); and mush (mersh). One often heard the "musk" in "muskmelon" and "muskrat" pronounced "mush" or "mersh." To some, sumac was "shoemake." "For" was commonly pronounced "fer"; a supposedly clever answer to the question, "what fer?," was "cat fur to make kitten britches."

One of the most common infractions of the rules of good English was the use of the double negative, such as: "You hain't seen nothin'." General use of "hain't" and "ain't" made them practically standard words. Many habitually bungled verb forms. They used "seen" for "saw" and almost invariably put "done" where "did" belonged. Very often "knowed" crowded "knew" out altogether. Case forms of pronouns tripped a great many, not a few of whom should have known better; one often heard such phrases as "with he and I" and "me and him went."

There was a strong tendency to maim present participles by prefixing an "a" and dropping the final "g". Such words became "a-comin'," "a-workin'," etc. The speech of a little group at school, children of the only Welsh family in the

201

district, stood out in marked contrast with that of other pupils because in every instance, when they used such words, they omitted the "a" prefix and sounded the "ing" very distinctly.

Following are a few of the more common words that made up local vocabularies. Although nearly all of them have been recognized and listed by lexicographers, some perhaps would not be generally accepted or approved as polite, refined English.

Brash (adj.), rash or impudent, also brittleness in timber; *breachy* (adj.), applied to a horse or a cow that could go over or through a fence; *britches* (n.), pants; *budget* (n.), valise, satchel, or package; *bulgine* (n.), locomotive or other steam engine; *crowbait* (n.), a decrepit horse; *dauncy* (adj.), indisposed, under par physically; *dingus* (n.), an unfamiliar object, a gadget or "gismo"; *doodad* or *doofunny* (n.), same as *dingus; drizzle* (n.), light rain; *dumbhead* (n.), a stupid fellow, an ignoramus; *Dutchy* (adj.), odd, slovenly, unstylish.

Fetch (v.), to bring; *fizzle* (n.), failure; *frogsticker* (n.), a dilapidated pocket knife; *galluses* (n.), suspenders for trousers; also four hills of corn tied together as the foundation for a shock; *galoot* (n.), an awkward or ignorant fellow; *gaumy* (adj.), smeared; *goo* (n.), anything sticky or messy; *green* (adj.), unsophisticated, inexperienced; *gum* (n.), either rubber, or a section of a large hollow log; *heaver* (n.), a horse with heaves; *hoof* (n.), one's foot— also used as a verb: to "hoof it" was to go on foot; *lickety-split* (adv.), moving fast; *lizard* (n.), sled for moving timber or stones; *mossback* (n.), an uncouth fellow, a dweller in a backwoods hinterland; *mouthy* (adj.), loud, over-talkative; *mudboat* (n.), same as lizard; *necktie* (n.), a

202

yoke to control a breachy animal; *pack* (v.), to carry; *peart* (adj.), bright, lively; *plug* (n.), an old horse; *polecat* (n.), a skunk, or a base, contemptible fellow.

Rain hen (n.), turtle dove; *rassle* (v.), wrestle; *rig* (n.), buggy or machine; *roof rabbit* (n.), cat; *saphead* (n.), a green or ignorant person; *shoat* (n.), a pig; *skedaddle* (v.), to flee or run; *snoot* (n.), one's nose; *soldier* (v.), to loaf or dally; *somers* (adv.), somewhere; *trotters* (n.), one's feet; *turnip* (n.), a cheap watch; *water brash* (n.), indigestion; *whale* (v.), to whip or flog; *whopper* (n.), a lie or anything large; *windy* (adj.), applied to a loud, boastful person; *yaller hammer* (n.) the flicker.

NINETEEN

North School, District No. 10

OUR TOWNSHIPS WERE DIVIDED into school districts, and a one-room schoolhouse was provided for each district. The building in each case occupied a small plot of ground near the center of the district. Schoolhouses throughout the region were all nearly the same in size and arrangement, many of them substantially built of brick.

Ours was North School, District No. 10, named for a family living near it (almost all rural schools were given their names that way.) The entrance was at the center of the south end. At the other end, in front of the blackboard that extended the full width of the building, was the teacher's desk. Each of the two sides was broken by a row of high windows covered by steel screens. Through the center, lengthwise of the room, was a wide aisle, unobstructed except by the big stove and a receptacle for fuel, both near the center of the room. The Board of Directors provided a generous supply of the best firewood for heating. We found it piled in neat, long ranks near the door when school began each fall. The upright pot-bellied stove could radiate a vast amount of heat; unfortunately, the heat was never properly distributed. Near the stove it was much too hot—a veritable torrid zone—while in far corners it was chilly on cold days. Pupils turned this to good account to overcome the tedium of lonely study at the remote desks they normally occupied. Several times daily, nearly all but those regularly seated in the well-heated zone would request permission to move temporarily nearer the source of heat when the weather was cold. In consequence the entire student body tended to be clustered near the center of the room during a considerable part of the time in winter, a disposition highly favorable to much whispered conversation and group fun that otherwise would have been impossible. A few teachers allowed themselves to be taken in by this pretended suffering in the far corners, but the wise ones simply opened drafts a little and allowed the heat to beat back the malingerers.

For quite a while it was customary to have two terms of school each year. The short spring term, which began in May, was provided especially for beginners and other younger children. It was always in charge of a woman teacher. I remember vividly my first teacher, a comely young woman who impressed me at the time as a paragon of feminine beauty and charm. The smell of lilacs always carries me back to that time. I see again my youthful classmates and that bright young teacher, on her desk a tall vase filled with fragrant purple lilacs. I liked the flowers and their perfume so much (we had no lilacs at home) that one of the boys brought me a vigorous shoot from his mother's garden. In our lawn it throve and grew into a large clump, its blossoms delighting us all each spring.

Soon after the county fair the fall-winter term began. Few but the younger children attended at the outset. Most of the larger boys and girls got started in October, after the

206

rush of fall farm work was over. In most rural schools this term was taught by a man, because it was generally felt that a woman might lack the physical strength necessary to curb misbehavior on the part of some of the larger boys. In a neighboring district a spunky young lady one year demanded, and was given a chance, to prove the fallacy of this notion. She came through with flying colors.

There were no grades as in modern schools. With one teacher and forty or more pupils ranging in age from six to sixteen, all in a single room, there could not well have been. The beginners started with a "primer," sole text for their first year. In the next higher class, the second-year group, were those who had advanced to the First Reader; the third-year group was in the Second Reader class; next in order came the Third Reader class. Those in the Fourth Reader studied practically all subjects in the school's curriculum. In his sixth year the pupil started in the Fifth Reader class. Here he was in the top echelon—there were no higher grades or classes. If one remained in school after his sixth year, as some did, because there was little else to do in winter, he went again into the Fifth Reader class and covered essentially the same ground as before except that, for older pupils who wished it, advanced instruction in arithmetic and algebra was available.

There were no formal graduations or promotions, no officially recognized flunking. When a pupil completed a school year in one reader, he advanced himself to the next higher. Qualified or not, he was usually allowed to remain. Naturally, some were beyond their depth much of the time. A few—not all of them actually dull—could not be interested in learning. Some in this group had fallen behind and never caught up because, for one reason or another, they had missed too many of the fundamentals. For most of the laggards schoolroom work was a bugbear.

In reading classes practically everyone read in a monotonous, unnatural singsong, especially when they were dealing with poetry. In expressing their own thoughts, all could speak forcefully and naturally; but in using voice modulations, accents, pauses, inflections, and emphases to give to the written word life and meaning and the feeling of the author, they failed dismally. This failure no doubt was due to precedent and bad example—younger pupils acquired the habit as a result of hearing older ones read, and it remained with them.

No one except a few of our teachers bothered to provide wholesome supplementary reading. It is appalling when one considers the aggregate amount of time that was wasted in country schools of that day, time that so easily could have been used to introduce impressionable young people more effectively to the rich, boundless fields of learning.

From time to time pupils brought in books of one kind or another. Many a blood-and-thunder novel did some of us borrow from classmates and read surreptitiously. If this had been generally known, it might have meant a very black mark against us from some in the community. To those good souls all novels were trash. (They were 100 per cent right, so far as some of that "literature" was concerned.) A few held that all imaginative works of literature, though they came from the pens of writers recognized by competent authorities as masters, were deadly pitfalls designed by the Prince of Darkness to entrap the unwary. Luckily, our clandestine haphazard reading included some good material, such as Grimm's Fairy Tales, the tales of

Hans Christian Andersen, a history of the Civil War, and a biography of P. T. Barnum.

I was kept from school until my brother attained school age. I must then have been a dreadful bore to the teacher, for I was several jumps ahead of other beginners: I knew the alphabet, could spell some simple words, and could read a little, having been taught by my father, who used newspapers as texts. We lived a mile and a half from school. Perhaps because he anticipated some thorny experiences for us in those long trips to and fro, Dad negotiated with a neighbor boy, a few years older, promising a reward for him if he would keep an eye on us and steer us away from trouble. We knew nothing of this until long afterward, but the arrangement unquestionably saved us from some hard knocks and brought us safely through situations that otherwise might have been rather unpleasant. In spite of this unsuspected guardianship, we fell into an adventure the very first day that, though we were wholly innocent of any intent to do wrong, brought us no little trouble. We were wearing bright new caps fresh from the store and new pants and waists that Mother had made expressly for our school wear. Crossing a bridge over a little creek on our way home, we went down with two other boys to look into the water. Immediately, we saw a school of small minnows. Into the water we went, wading in pursuit of them. In the excitement of the moment, finding that we could not catch them in our hands, we pressed our caps into service as seines. We succeeded in dipping up half a dozen of the little fish, which we transferred to our dinner pail with half a gallon of water. Then, proud and elated, we trudged homeward, bearing the living trophies of our prowess. We must have made an exceedingly disreputable appearance, our new caps wet and twisted out of shape, our pants and waists dirty and bedraggled, and our hands, faces, and hair smeared with mud. It was just too much for Mother, who had started us off so proudly that morning. She promptly administered to each of us a sound smacking, right on the seats of our wet pants. That paddling was meant to be impressive. It was—we remembered it a long time.

At school that first morning, we novices saw one of the boys sitting on a desk, his feet resting on the attached seat. When the bell rang, we naïvely seated ourselves side by side in the same manner on a desk, feet on the seat, our books, pencils, etc., in our hands, our dinner pail between us. This evoked a great deal of snickering and tittering until a more sophisticated lad nudged us and indicated that we were supposed to sit in the seat, not on the desk. That was Lesson No. 1.

At first we had an old schoolhouse with weatherbeaten walls, dingy windows and door, and rickety desks deeply carved and defaced by the jackknives of preceding groups of boys. When time came for school to "take up," the teacher, standing at the door, rang a brass hand bell.

In the summer after our second year, a new building was erected on the site of the old, essentially the same as the latter except that it had, instead of a black-painted area of wall plaster, a large slate blackboard. We noted also an elaborate chart designed to make grammar plain and simple. There was a big new bell in a small belfry above the entrance, with a long rope extending down through the ceiling. This bell thereafter called us from play to books. Most of the seats, bright and new, were made to accommodate two pupils each. It was the custom for each

210

pupil to choose a special friend as his seatmate. Together, the first day, they decided which seat they would occupy. These arrangements were seldom lasting. Generally the teacher had to separate several pairs within a few days because of excessive whispering and cutting up. Not infrequently, a quarrel would break up a twosome. The sexes were separated as in church, girls being seated at the right, boys, at the left.

About half of our teachers came from nearby districts, half from outside areas (one, from a county many miles away). One or two had some college training; some had not progressed beyond high school. Several had no formal education beyond that acquired in district schools such as ours. The latter, incidentally, were among the best of the lot.

Teachers were employed by the Township Board of Directors, who investigated each applicant's personal character, scrutinized his letters of recommendation, and made sure that he had a Teacher's Certificate. Often, the applicant was known personally to some or all members of the board. All prospective teachers were required by law to pass an examination given by the County Board of School Examiners, covering all subjects taught in rural schools, plus pedagogy.

For young men, teaching was often a steppingstone toward training for a career in business or a profession. For a salary that averaged about forty dollars per month, a teacher was expected not only to provide instruction but to assume responsibility for the children at play and in their going to and returning from school. From first to last, he served, willy-nilly, as a sort of exemplar and mentor in the community, all his habits, words, and deeds constantly subject to the closest scrutiny and appraisal. On top of all that, he had to serve as janitor of the building.

One of our women teachers unfortunately got off on the wrong foot at the very first. She tried earnestly, but she went about everything so tactlessly that within a week she had antagonized about half of the pupils in her charge. Thereafter, they took special delight in making trouble for her. One of them came one day with this doggerel, which he zestfully recited to everyone who would listen:

> *O Lord of love,*
> *Come down from above*
> *And bless us poor scholars;*
> *We hired a fool*
> *To teach our school,*
> *And paid her forty dollars.*

Probably the most difficult problem of the teacher who had the winter term in a rural school was that of discipline —especially holding in line the big boys who attended. This problem appears to have been created and kept alive largely by tradition. The boys heard from grown men—in a few instances their own fathers—how, in their day, they had defied their teachers, cowed them into yielding important disciplinary points, and, in some cases, "licked" them in physical encounters. These men had little respect for education; they held that learning beyond the ability to read passably well, to write a legible hand, and to do the elementary "ciphering" incident to farming was superfluous and silly.

It was common practice for a teacher to bring in, about the first day, a long, lithe switch cut from a tree. This was

212

kept within convenient reach, where all could see it. Some modern pedagogues quite likely would take a very dim view of that practice. Nevertheless, under circumstances that then prevailed, that whip had a pronounced salutary effect.

If a teacher showed a disposition to be perfectly fair, coupled with an unmistakable determination to brook no infraction of any rule of good conduct, he soon had the burly boys eating out of his hand. At times they even vied with one another for the privilege of carrying wood and helping with janitorial duties. On the other hand, if a teacher failed to show firmness at the outset, if he yielded or compromised, the initiative was likely to pass quickly from him. Regaining the mastery then was not easy, but some did it by soundly whaling a young bully or two. There were no namby-pamby notions about corporal punishment in the district. If a teacher whipped a boy for misconduct, it was generally felt that he was merely doing his duty. Legal actions against teachers in such cases were unknown.

When we boys started to school, our parents made it quite clear that they expected us to obey the teacher under all circumstances. "If you misbehave and get a whipping in school," said Dad, "you will get another at home." This had weight with us; we knew that they were not just idle words. We carefully observed the letter and the spirit of school rules and during our first years got along wonderfully well.

Then, one day, in spite of good intentions, we found ourselves in real trouble. Some of the larger boys, eager for excitement and diversion, contrived to get a fight started among younger boys at the noon recess. Soon four boys, including my brother and myself, were involved in

213

the melee. When we assembled for the afternoon session the teacher, noting black eyes, scratched faces, and bloody noses, tanned all four of us. We bribed our sister to keep the matter secret at home, promising to pay well for her silence as soon as we got the money. She kept her part of the agreement for two or three years. Then, angered by something that we had done and remembering that we had never made good our promise, she told Dad the whole story. Luckily, something like the statute of limitations worked in our favor—we never got that home licking.

Outstanding among our teachers was a young man with wavy red hair, a face peppered with freckles, and an exceptionally agreeable personality. He quickly won the friendship and esteem of all in the community, including his pupils. He was a master teacher, with a thorough knowledge of all subjects taught. Right in the beginning all in school, including potential troublemakers, knew without a word being spoken on the subject that he would stand for no tomfoolery. (It was he, by the way, who gave my brother and me the only larruping we ever got in school. This did not lessen in any degree our liking for him, although I have always felt that the instigators of the fracas that noon deserved some attention.) From teaching he went into the business world, eventually becoming owner of a successful business enterprise.

Another well-liked young man was an exceptionally competent instructor who joined heartily in nearly all of our play. This alone would have been enough to win for him warm friends and supporters. He was firm enough to maintain good order and tactful enough to promote enthusiastic co-operation, even in the school's singing. For years after he gave up teaching, he practiced law success-

214

fully. He bequeathed to the town a public historical museum, providing in his will for its maintenance.

As happens today, many little boys attending spring sessions fell in love with their pretty girl teachers, and, not infrequently, some older girls became enamored of the male teachers. The men were not always aware of this; but when they knew, they managed so tactfully that there was no embarrassment, and no hearts were broken. In a few instances puppy love between boys and girls blossomed and grew, finally, when they had grown up, culminating in marriage. More often, however, the eyes of the boys would be taken by girls in neighboring districts, another county, or even in some far removed town. Local girls were sought out by boys from other schools, near and far. This seems to bear out the adage about the grass appearing greener on the other side of the fence.

We had a well and an iron pump a few steps from the entrance door, with a rusty tin cup on a wire hook. A few pupils had private drinking cups, but the majority saw nothing at all wrong about using the common cup. If one were thirsty (or tired of schoolroom routine), he asked permission to get a drink. Carrying his slate-cleaning sponge, presumably dry, he went out, and we would hear the pump creak long enough to bring up a barrel or two of water. Several minutes later (kids usually stayed out as long as they felt would be safe), his thirst assuaged, he came noisily in and resumed his studies. Some of the more expert time-killers would save the sponge for another trip to the well a little later.

The greatest possible amount of play was crowded into our three daily recess periods. Some swallowed their midday meals almost whole to avoid any unnecessary waste of the noon hour. We never tired of playing ball, using at times homemade yarn balls with rubber cores. The yarn generally came from raveled stockings. One could buy molded solid rubber balls, but they were of inferior quality —they split too easily. Now and then, someone would get a hunk of solid rubber, known as "car rubber," probably because its original use had been as a cushion in a railway car. He would whittle from it a rough sphere about two inches in diameter. Such balls were very tough and so resilient that a good batter could hit them several hundred feet. We liked them especially in what we called "Rounder," a game somewhat similar to the present baseball. Though quite hard, we caught them, even on the fly, sometimes wearing only thin canvas gloves, sometimes barehanded. We knew nothing of padded gloves. Bats were all homemade, whittled out of oak or ash. Girls frequently joined in ball games, some of them being excellent players.

Many times we "rasseled" hats. The players, standing in a circle, threw their hats into the ring. One, blindfolded, shuffled the headgear, then came up with a hat in each hand. The owners of the hats wrestled until one was thrown. The shuffling was not always strictly honest; the shuffler might peek enough to get the hats of two boys well matched in size and wrestling ability in order to be sure of an exciting, worthwhile contest. From this game, in all probability, came the expression so much used in political circles—the man who becomes a candidate for an office is said to "throw his hat into the ring."

"Shinny," a rough winter game similar to hockey, was great fun. Sometimes our puck was a wooden ball, sometimes a rock or a battered tin can. Our clubs were cut from tree branches with a natural bend or from the bent bows of old buggy tops. When the pond at the tile mill froze over, we transferred this game to the ice, playing on skates. At other times, we played on the frozen ground.

With a single exception, all of our games were intramural. That exception was a football game played in the snow one December day against a neighboring school's team. The only remembered feature of this contest is that it provided the setting for my first meeting with Cletus Wright, who, through all the succeeding years, has been among my cherished friends. A keen lover of sports and possessor of an extraordinary sense of the dramatic, he came that day from a district several miles away to see the game. He arrived with a wild-West flourish, his snow-white horse running at top speed, drawing a fantastic homemade sleigh that slithered and careened crazily in a white cloud of snow. A few years later he and I became classmates and roommates in college, living, working, and playing together through several years.

None of us at North school had ever seen skating shoes. A few had heel-plate skates with knobs that fitted into steel heel-plate sockets set into boots or shoes. They were fastened with straps. Most of us preferred the "Winslow" or "Barney and Berry" adjustable clamp-type skates that, at the push of a lever, gripped heels and soles and required no straps. The mill pond and Dog Creek, which flowed by it, afforded excellent ice for skating; many a noon hour did we spend on one or the other. The creek ran through our farm and, about a half mile from the house, crossed our regular route to school. Daily, when the ice was good and free from snow, we skated to and from school.

Often in early spring we played a rural version of the cops-and-robbers game. First we built a "jail" in the corner of a rail fence. Then we elected a sheriff. All players except the sheriff assumed the roles of horse thieves. It never occurred to any of us that it was a bit out of the ordinary for thieves of such a despised order to make provision, as we did, for having themselves brought to book. The officer promptly set to work to round up the lawbreakers. This was a tough assignment because hideouts were plentiful, and most of the pursued were as fleet of foot as the pursuer. If one could outsmart the law or wriggle free after being apprehended, he might remain at liberty quite a while. But once the sheriff led a desperado to the jail, thrust him through the door and called out "click-clock," the rules decreed that he had to remain a prisoner. The poet sang, "Stone walls do not a prison make," but for us, sticks laid on the ground made prison walls, bars, and door. Rules of the game made escape by breach of the walls impossible. However, if the sheriff forgot to kick into place the stick forming the door and then lock it with a resounding "click-clock," he would invariably find all the birds flown when he returned to the clink. This was an exciting game, but it required a great deal of leg work on the part of the sheriff. Furthermore, if it was not a moderately warm day, thieves in hiding or in durance vile might get pretty chilly.

One winter someone in school read a magazine advertisement offering a marvelous opportunity for youthful salesmen to earn beautiful, useful premiums. He sent in his name and by return mail received a list of enticing premiums, order blanks, and samples of extraordinarily ornate cards. Not calling cards, mind you, not business cards, but "hidden name" cards! The manufacturer called them that, and we assumed that he knew what he was talking about.

The cards, extremely gaudy, came in an assortment of sizes and shapes, with deeply scalloped, tinted edges. For

218

the printing of his name, one could choose from several styles of highly ornamental 18-point blackface type or from a number of ornate script styles, all embellished by an amazing lot of flourishes and curlicues. A multicolored, hinged paper cutout, with garishly lithographed and varnished floral or bird-and-angel decorations in considerable variety, covered the name on each card. The cards caught on quickly. After the first two or three had placed orders, the salesman had only to write orders in rapid succession and collect the money. Collection in some instances required a great deal of time and patience, but eventually everyone paid up. In due time the cards were delivered as ordered, together with the salesman's reward for faithful, earnest effort—a small trumpet made of bright tin with two or three erratic keys. The cards were distributed to customers and then a general exchange took place, each trading one of his own for the name card of a friend.

From time to time after that, others of our classmates in the role of salesmen called on us, offering such necessities as chewing gum, small squares of scented chalk posing as sachet perfume, and small, embossed, silver-washed souvenir spoons. After disposing of their quotas of merchandise, all of these salesmen received premium awards such as Ingersoll watches, trick puzzles, pasteboard cameras, and, in one case, an asbestos tobacco pipe formed and colored to look like a cigar.

Most of the boys and a few of the girls acquired nicknames. In a few cases the origin of these names was obvious, but many seemed to have been applied without any reason whatever. We had Peeky, Buck, Stump, Dentist, Flippy, Cock Robin, Seeky, Welshy, Dog, Egg, Shep, Tug, Doc, Grinny, Felix, Bill Nye, Button, Fox, Queasy, Brick, Cricket, Scaley, Dime, and Cookus.

Poor Cookus had a hard time of it in school. He couldn't get interested in books, and never saw much sense in any part of school except playground activity. One day, to keep himself occupied, he tore the red felt binding from his slate. (Everybody had a slate, supposed to be cleaned by wiping with a damp sponge, but in too many cases "cleaned" by spitting on it and wiping it with a rag.) Finding that the felt when wet would give up its dye, Cookus swabbed his face with it, and his visage became as red as that of any Indian that ever prowled the forest. Taking note of the lad's odd look, the teacher brought him forward and made him stand, grinning shame-facedly, where all could see him. For several minutes there were long, loud guffaws at his expense.

Arrangements were being made one time for a little program of recitations, songs, and so forth, to be put on in lieu of regular classes the following Friday afternoon. All but Cookus entered wholeheartedly into the plan. He held back stubbornly, offering every imaginable sort of excuse for non-participation until at last he saw that there simply was no getting out of the thing. When his name was called on the appointed day, he stepped to the plat-form, made his best bow, and treated us to the following:

> *Seen a rat run up the wall;*
> *Seen its tail and that was all.*

Although devoid of interest in books and school tasks, he could recognize and name at sight all of the birds indigenous to the area and could imitate their calls with surprising fidelity. He was extraordinarily fleet of foot and was the best skater among us. He could throw a

220

stone with the precision of a sharpshooting rifleman, and he threw with incredible force. Surely he would have become a notable baseball figure if he had been given a chance.

Cookus and I once figured in a bit of schoolday drama that I have never ceased to regret. He was about twelve, I about ten at the time. We were walking in a little group headed homeward when Cookus announced his determination to kiss one of the girls. No sooner were the words out of his mouth than she was off, running like a fawn, he, close at her heels. Rooting enthusiastically for Cookus, I began swinging my dinner pail wildly round and round. About the fourth round, a sharp edge of the pail smashed into the face of "Egg" Evans who, altogether unnoticed, was running closely behind me. A long, deep gash was cut in his cheek that left a permanent ugly scar.

After they left school, two of the boys among my classmates signed up with Uncle Sam for a hitch in the navy. One was Cookus. He had been aboard ship as a stoker only a few months when his trunk was shipped home. Laconic official word came to the family that he had been lost at sea; no details whatever were given. As the country was at peace, the fate that actually befell him is a mystery to this day.

With the exception of two or three who lived close to the school, all of us, including teachers, carried our lunches every day. The majority lived a mile to a mile and a half distant, and all walked, regardless of mud, ice, and drifted snow, trudging over the roads, some macadamized, some plain dirt, or hiking across fields and climbing fences. When it rained, some fathers drove their children to school in curtained carriages. Once in a while, to the envy of most others, a boy got to ride with his dad on a horse, both wrapped in rubber coats.

Daily, each of us carried to school, and ate, five or six apples, the best to be found in the home store. Two or three. times each week, someone brought in a big lot of sorghum taffy or a full peck of popped or parched corn. The aroma of the corn pervaded the room immediately, whetting appetites and centering attention upon the girl or boy who had brought it. Providing these treats had become an established custom, and each of us who cared for the full approbation of schoolmates felt an obligation to contribute from time to time.

Occasionally, instead of a "literary" program for Friday afternoon relaxation, we voted for a "spelldown" or a "ciphering match." For either contest two leaders, closely matched as to ability, were named. The teacher, with a finger between the pages of a closed book, would ask each leader to guess the page so marked. The one who guessed nearest was entitled to first choice in naming members of his team. Soon, all in the school were aligned on one side or the other.

In a spelling match members of one team stood along one wall, their opponents along the opposite wall. The teacher pronounced words from a spelling book, starting with comparatively easy ones and gradually working into more difficult lists. The first word went to the first in line on one side, the next, to the second, and so on down the line until someone missed. He was obliged to take his seat when someone on the opposing team, often a specially selected "trapper," spelled the word correctly. The spelling continued, along one line and then the other, each contestant taking his seat after being spelled down. The one who remained standing when all the others were down won the contest for his side and made himself for the time being the champion speller. Two girls, cousins, were so good at spelling that nearly always one or the other was the winner at a spelldown. They took the matter so seriously that, almost without fail, the defeated one went to her seat in tears. Spelling contests, coupled with the strong emphasis put on spelling as a subject of study, made a few country pupils excellent spellers.

Most pupils seemed to enjoy ciphering matches more than spelldowns, probably because the former tended to be a little more exciting and because, as usually happened, nearly all of the real work would be done by half a dozen of the school's brightest students. The best were always first choices of leaders in either of these contests. One could have gauged quite accurately the relative abilities by noting the order in which pupils were selected.

After two ciphering teams had been formed, each leader named someone from his side to compete in the initial contest, beginning with the least competent. The one who came up first with the correct answer to a problem was the winner. At times a majority of wins in several trials was required. Each loser returned to his seat, and the leader named someone to replace him. This continued, each leader going up the scale of age and ability in naming contestants and the teacher propounding more and more difficult problems, until all on one side or the other had been "ciphered down." The last at the board, undefeated, was the champion, and his side was winner of the match. Quite often, the champion was one of the leaders. Several in the school could perform intricate arithmetical operations with amazing speed and accuracy.

In regular spelling classes all stood in line, the right hand end of the line being the "head," the other end, the "foot." The first word went to the pupil at the head. If he spelled it correctly the next word went to the second pupil and so on down the line, then back to the head. If one missed a word, the teacher called "next" and the next pupil gave it a try. If he got it right, he exchanged places with the one who had first missed it.

Thus it was possible, if one had luck and spelling ability, to move all the way from the foot to the head of the class, either by degrees or by a single correctly spelled word. A "headmark" went to the pupil who stood at the head at the end of a recitation; he had to start at the foot the following day, all others retaining their relative positions. A daily record of headmarks in each class was kept. A number of teachers, aiming at stimulating all to their best efforts, credited the marks opposite pupils' names on a chart kept in daily view.

Bowing to precedent of long standing, our teachers always provided, at their own expense, a treat for all in school at Christmas. They kept all preparations as secret as possible and often went to great lengths, short of outright lying, to make everyone believe, even up to the last minute, that nothing at all would be forthcoming. Regardless of what might be going on, just at the proper moment about midafternoon of the last school day before Christmas, smiling, white-whiskered Santa Claus, decked out in a red cap and a long, fur-trimmed red coat (which partially hid his blue overalls and high rubber boots), burst open the door and strode noisily into the room. He carried several big bags, and sleighbells jingled loudly at every step. Generally, he refrained from talking, knowing that his voice would identify him. The presents, usually a bag of candy, a bag of nuts, an orange, and a ruler or pencil, were distributed amid great, happy excitement.

224

One time, Santa, distributing gifts, came upon a little boy of preschool age who had "happened" to come as a visitor. As kindly St. Nick passed out gifts to him, the lad piped out: "We got three more kids to home." With a great belly laugh, Santa produced gifts for each of the absentees (teachers always provided enough to make sure that no one would be missed).

Santa came one Christmas and brought his "son," a man who lived near the school. This four-footer, known everywhere as "Shorty," was a born clown. His antics kept everyone laughing while he and his "dad" distributed gifts. Smoking a corncob pipe, he walked to the platform just as they were ready to leave and announced that he would "speak a piece." Then, with a sweeping bow, he gave us this:

When I was a little boy, my mother kept me in.
Now, I'm a big boy, fit to serve the king.
I can have a musket; I can smoke a pipe [flourishing pipe];
I can kiss the pretty girls at ten o'clock at night.

Shorty came on another occasion, a Friday afternoon that had been set aside for entertainment. He brought with him a parlor organ and two neighborhood fiddlers. Without ado, they started the show, the fiddlers scraping away at their instruments, both tapping their feet in time, and Shorty "chording" at the organ. Their music, all lively old-time tunes played one after another in rapid succession, was accompanied by superb, masterful clowning, making the performance, from beginning to end, a delight.

Occasionally, a parent or other resident of the community would come for an afternoon visit. The teacher always tried to conduct classes as usual and to keep things going naturally and normally. Invariably, these efforts were nullified in large measure by his charges. The visitor's arrival was taken by some as a signal to begin showing off. They would self-consciously simulate intense interest in their studies, breaking this commendable application by frequent side glances at the visitor. Later, they were likely to engage in long whispered conversations, feeling sure that the teacher would hesitate to reprove them. The cutups would seize opportunities to hurl wads of paper or pieces of chalk. Requests to leave the room or to speak to someone were sure to be twice as numerous as when no outsider was present. Reciting in classes, many would display unwonted ignorance and stupidity.

Now and then, a district school somewhere about us would name a night for a spelling bee and invite all who wished to come. Often the house would be jammed to capacity. The contests were exciting because most of the real spellers came from near and far, and they came for blood. Equally popular were "literaries," programs commonly made up of recitations, some group singing, maybe an oration, and always a debate, on a subject announced weeks in advance. The debates were informative and entertaining because the participants were well-informed young men—most of them teachers—who had forensic ability and knew how to argue logically and convincingly.

The crowning event of the school year came toward mid-April, at the end of the winter term. This was the "Last Day," a traditional gala in rural schools of the area. On this day, long anticipated, the girls and boys wore their very best. All were on their best behavior, trying to make themselves and the school appear in the most favorable light. The teacher, fully aware that embarrassing situations could arise, was hoping for the best and praying

226

that somehow he would get through the difficulties of the day successfully. A little before noon, fathers and mothers began driving into the yard, bringing big, heavy baskets of food. They found seats, and with lively interest watched their young hopefuls at their studies and recitations, speaking meanwhile to each other and to neighbors, sometimes in whispers, sometimes in tones that carried to all parts of the room. Visitors from other districts also were showing up, young men and women much more interested in the social aspects of the occasion and the prospective bounteous dinner than in any evidence of scholastic progress that might appear. Usually, a minister was on hand, smiling benignly and shaking hands amiably with everybody he could reach.

When classes were dismissed at noon all of the pupils and all of the men filed from the room, leaving the good matrons of the district in undisputed charge. The smaller children, trying (though often quite vainly) to keep their clothes clean, ran about in little groups, talking, giggling, and playing simple games. The larger boys of the school and the young men visitors whiled away the waiting period by engaging in such athletic contests as the running broad jump, the standing broad jump, and the hop, skip, and jump. Their best coats draped over the board fence, they strained nerve and sinew in these trials, getting shoes and pants smeared with mud as, at the end of each powerful spring, they landed in the soft, well-churned soil.

At length the bell sounded the dinner call and all crowded into the building. There, spread out over cloth-draped boards laid across desks, was a magnificent display of food. Included were roasted and fried chicken, savory beef cooked in a variety of styles, stuffed and pickled eggs, baked hams, baked beans in many shades and consistencies, pickles and relishes of every variety, more than a score of cakes and as many pies in mouth-watering array, quantities of cookies, tarts, rolls, jams, jellies, candy, nuts, fruit—just about every item of food that one could name. The teacher, as master of ceremonies, tapped a plate with a spoon until the chatter and hum of voices subsided, then asked the minister to say grace. Everybody fell to after that. The smaller children, big napkins over their fronts, were helped by their mothers. All others served themselves cafeteria fashion, moving from table to table, all talking merrily as they ate. One heard lively banter, jokes, and bon mots galore—all funny, but nothing meant to be malicious or disparaging. There is a question whether one could get together any group of everyday people whose quick, clever witticisms, spontaneous gags, laughter-provoking retorts, and pranks would surpass those of an assemblage of farm folks who know one another well.

Within a short time, groups of food-stuffed, giggling children, holding in their hands pieces of cake and other tidbits, began pushing their way outside, eager to resume their play. Some fast eaters among the men, having gulped their fill, also worked their way out, to get away from the congestion and to enjoy their after-dinner smoke or chew of tobacco. Those remaining inside, taking advantage of the less restricted elbow room, now moved about more, sampling all tempting viands in sight. At last, everyone having crowded digestive capacities nearly to the limit, the tables (still loaded with more than enough food to satisfy another company of like size) were left to the attention of the good ladies. Each collected her dishes, table linens, and other belongings, and packed them, with remnants of the repast, into her basket.

The afternoon was given over to a program in which

228

each pupil participated in some way, to the end that something of the school's achievements might be demonstrated and that each parent might find in the performance some grain of nourishment for his or her personal pride. After this, visitors and parents were invited to make "remarks." The minister would speak briefly, making his words as humorous and as complimentary as he could. A few of the parents might make short speeches, lauding the teacher's efforts and seeking to encourage the pupils to continue in the pursuit of knowledge. The teacher might say a few words, telling of the satisfaction he had derived from association with the school and the community. Then, seated at his desk, with a quantity of small coins before him—cash from his own pocket—he began calling the roll, requesting that the children come forward as their names were called. As each in turn stood before his desk, he gave him the money due for the headmarks he had earned, and with a friendly smile, bade him good-by.

District No. 10 schoolhouse was abandoned and torn down years ago. Looking back and considering the many far-reaching changes wrought by time, the poem, "Forty Years Ago," in our McGuffey Fifth Reader, comes to mind. The first stanza ran:

> I've wandered to the village, Tom,
> I've sat beneath the tree,
> Upon the school house playground,
> That sheltered you and me;
> But none were left to greet me, Tom,
> And few were left to know,
> Who played with me upon the green,
> Just forty years ago.

How many times, as boys and girls in school, we read that poem! But, being young, we naturally read it always objectively and impersonally. The words to us were those of some queer fellow, extremely old and doddering. For us, such a retrospective look, after such an incredibly long span of time as forty years, seemed almost impossibly remote. Now, we can discern real meaning in the poet's nostalgic words and appreciate how true and faithful was the picture he painted.

In 1892 the Ohio legislature enacted a measure known as the Boxwell Law, which provided that pupils from rural schools, successful in examinations prepared under provisions of the law, could be enrolled in urban high schools, their tuition paid from educational funds of home districts. The number taking advantage of the opportunity thus afforded was relatively small. Over a long period prior to that time, a few young men and women educated in one-room district schools of the region became teachers in such schools, some with supplementary education in one of the several normal schools, some without.

A handful of young men, some of whom had been both pupils and teachers in district schools, moved up through educational facilities available to them to enter business or a profession. Few achieved special fame, although one, from the little rural school that my father attended, was for a long time widely and favorably known as editor of an important city newspaper.

In most cases the grounds once occupied by district schools have long since reverted to the farms of which they were originally a part. Nothing remains to indicate that they were ever sites of school buildings. Farm children now ride in busses to centralized grade and high schools.

A fair proportion of these youngsters go on to college. The girls enroll in much the same courses that their city cousins take. Some of the boys prepare themselves for business or professional careers. Others major in agriculture; but oddly enough, few of the latter actually return to the family farm. They become college professors, government officials, county agricultural agents, managers of farm estates, specialists in industry, or salesmen for manufacturers of products for farm use.

TWENTY

The Fair

YEAR AFTER YEAR, IN THE second week of September the county fair was held. Very often, too, this was the week when corn-cutting should have been started. Getting the corn into shock was regarded as an important job, but few ever allowed it to shut them out altogether from the pleasures of the fair, which was the outstanding event of the year to nearly all in the county and to many in neighboring counties, both ruralites and urbanites.

The fair opened Monday morning and closed Friday night. There were no night sessions for many years; now they are commonplace and popular everywhere. In spite of frowns and protests from some who considered it a desecration of the Sabbath, hundreds of men and boys visited the fairground in Van Wert on Sunday afternoon. They watched concessionaires set up stands and tents. They sized up horses, cattle, and pigs as they trickled in to compete later for prizes. They sat on the board fence surrounding the elliptical racetrack and made jokes as trainers exercised trotters, pacers, or runners; or as local owners of classy horse-and-buggy rigs got on the track for a few rounds in the limelight.

Attendance on Monday was always light; but early on Tuesday, a big trek would begin, horse-drawn vehicles, bicycles, and occasionally pedestrians, moving in on all roads leading to the fairground. As the day advanced, the traffic increased steadily until, toward noon, it might become almost a solid procession. Some families piled into lumbering, springless farm wagons drawn at a walk by stalwart teams of horses, the passengers sitting on boards laid across the wagon box, their feet in hay or straw on the floor. Other families rode in two-seated carriages—nearly all of them with a fringe around the top. Some of these conveyances were drawn by teams of big work horses, some by single, light-footed carriage horses. At times numbers of people from farms and towns miles to the north, south, east, or west came by train.

Young blades, some with their best girls beside them, traveled in buggies whose shiny paint had been newly washed and polished for the big week, their fast-stepping horses well groomed and smartly caparisoned in neat black harness. A few of the ultra-stylish among them drove with white lines. All of these fellows delighted in pulling out of the procession from time to time, passing half a dozen or so slower vehicles and then cutting in where they could continue at a fast pace. To a casual observer it might have appeared that each was bent upon making everyone believe that he was a most important personage, whose time was of priceless value.

Toward evening each day the homeward movement would begin and would continue until after nightfall. The homebound traffic was rarely as heavy as that moving toward the fair at peak periods; the vehicles somehow became more evenly and more thinly distributed. The grinding of the wheels and the tramp of the horses' feet soon reduced top layers of roadways to a fine powder. Barring rain, this powder, near the end of the week, might be two or three inches deep over the dirt roads. Throughout the daylight period and far into the night, a pall of dust hung thick and heavy over the countryside, settling upon vehicles, their occupants, the horses, roadside fences, porches, crops in the fields, and even sifting into houses.

Few farm families ever thought of going to the fair without a basket dinner. At noon the clan gathered at the family wagon or carriage, and the edibles, the choicest of fare in generous quantities, were placed on a tablecloth spread over the cleanest grassy spot that could be found. Rarely was the best spot available really clean or ideally shaded. Dust lay thick everywhere, and bits of paper, melon rinds, banana peels, and cobs from which horses had eaten their lunches littered the ground. Swarms of flies persisted in efforts to get at the food. The ground was pretty well policed but it was not uncommon for sneak thieves to raid a farmer's wagon or carriage. Then, members of the family, coming together for the noon meal, would be discomfited and chagrined to find basket and contents gone without a trace.

The best horses, cattle, sheep, hogs, and poultry in the county were exhibited. There were displays of amazingly tall corn, excellent wheat and other small grains, tempting fruits, and perfect vegetables. Schools brought in select work done by their pupils. Of special interest to women were needlework, baked goods, jams, jellies, and canned fruit. Paintings and sketches in various media were displayed by local artists. Dealers and salesmen were in daily attendance, displaying and demonstrating farm machinery, musical instruments, household appliances, and furniture. All exhibits, each in its particular class, were in competition for prizes.

There was a full program of horse racing every afternoon. The majority of visitors were content to watch at intervals from free vantage points along the fence surrounding the track. Others preferred to pay a quarter a head for seats in the grandstand from which, throughout the afternoon, they had a good view of all parts of the track. Excitement at times reached a high pitch. There was a little betting, but it was on a much smaller scale than has been common in later times.

Daily, inside the track and directly in front of the grandstand, the balloonist and his helpers started inflating the big balloon about midafternoon. The operation was timed so that the ascension would come late in the day, in order that the crowd might be held on the grounds as long as possible. The balloon was supported by a rope stretched between two trees. In the ground, beneath the mouth of the big bag, was a pit in which, after ropes and sandbags had been properly arranged, a fire was kindled. The aeronaut or an assistant fed light wood and kerosene into the flames, using a tin cup to throw in the oil.

When the hot gases had inflated the balloon somewhat and it began to pull noticeably at its moorings, volunteers chosen from bystanders moved in and, gripping it at the bottom hem, held it in place while more oil was thrown into the fire. At last, when it was fully inflated, there was a cry: "Let 'er go!" Men and boys holding the bottom released their grip, the ropes were all slipped free, and the balloon dramatically started skyward. As it rose, the aeronaut, dressed in tights and seated in a trapeze below

236

the parachute, waved gaily to the spectators. Up soared the buoyant bag, higher and higher, until it appeared no larger than an egg and the swaying figure of the man a mere speck against the sky.

At the proper moment he "cut loose," and spectators held their breath until the parachute opened gracefully and began wafting its passenger gently earthward. Soon the balloon, weighted at the top, turned bottomside up, and dark, smoky gases poured from its mouth as from a volcano. Shortly, it collapsed into a limp rag, then fell swiftly, streaming and waving like a great dark banner, trailed by wispy smoke.

One year the ascension was made by a woman who made beautiful, spectacular flights. The possibilities for accidents in connection with all those performances were numerous, but our balloonists all came through unscathed. Amazingly enough, no trouble ever came from their careless handling of kerosene and fire. They always made their landings on open ground, rarely more than a mile or two from the point of ascent.

The visitor found at the fair a variety of catchpenny amusement devices. There was a striking machine with a wire-guided ball that a brawny customer, armed with a maul, could send high into the air, the top point of its flight measured on a painted scale. Several spots could be found where one might test his skill at throwing baseballs and maybe win as a prize a cheap black cigar or two. If one had a yen for pitching wooden rings, he could readily find a stand where, by very skillful tossing, he might win a cane, a knife, an umbrella, or even a doll.

Always at a spot where crowds tended to move in greatest numbers a glib talker stood on a platform with

237

a supply of buggy whips, making them crack like pistols and exchanging them rapidly for coin of the realm. Not far away one could find other pitchmen at drop-doors of their wagons, vending marvelous Indian cure-all medicines or lines of merchandise, often very gaudy and all of the most doubtful and ephemeral value. On every hand were peripatetic hawkers, setting up a great din in their efforts to attract customers. They offered for sale toy riding whips, gold-washed "jewelry," ice cream, candy, hot dogs, popcorn balls, rubber balloons, squawkers, and even live chameleons. It may well be that descendants of this fair-following gentry have become the hard-sell hucksters and spielers who now contribute so many obnoxious features to radio and television programs.

A merry-go-round and a Ferris wheel ran all day for the entertainment of children, young and old. It is a strange fact that boys and girls appear never to tire of riding these devices. Each crop of youngsters seems to enjoy them fully as much as did their parents, not to mention their grandparents, before them. The carousel's mechanical organ was equipped to play just one currently popular tune. It blared this forth over and over in loud, tinny tones, throughout each day of the fair. As a result, visitors might seem to hear the music hours after they had left the grounds, sometimes several days after the fair had ended.

The midway boasted several tent shows with long-winded barkers, featuring mediocre vaudeville acts, sleight-of-hand performances, wrestling, "world wonders," and freaks—nearly all brazen fakes. Generally, performers and attendants were frowzy individuals of obviously questionable character. One year we saw a show that stood out because it was honest and aboveboard. The performers were ballyhooed as "Bohemian glass blowers." They may or may not have been actual Bohemians, but they were such skilled artisans that it was both instructive and fascinating to watch them blow glass of many colors, fashioning it into ornaments and trinkets of extraordinary beauty and delicacy.

Everywhere one saw swains walking happily, hand-in-hand with their lady loves, all dressed in their fashionable best, all seeking to make the best possible impression and to cram into each day, each hour, the utmost in pleasure and entertainment. A carnival spirit prevailed, and everyone had a wonderful time, in spite of milling crowds, heat, flies, dust, and feet that ached and burned and throbbed in protest against endless walking and standing.

Year after year, a traveling stock company arranged its schedule to appear at the theater in town nightly during fair week. Usually, a different play was presented each night. The promise from Shakespeare's *As You Like It*, painted on the curtain, " . . . tongues in trees, books in the running brooks, sermons in stones, and good in everything," was rarely made good in full measure. But those who filled the house to see the plays, none too critical of the drama or the Thespian art presented for their edification, found pleasure in the performances, hammy though they were, and enjoyed the music furnished by a local orchestra. For many, a night at the theater rounded out a full day of pleasant, diverting activities, long to be remembered.

To not a few men and boys who, at the end of fair week, had to don overalls and sally forth to cut corn, the toilsome task came as a severe letdown, a painful descent from a plateau of pleasure to a morass of misery.

TWENTY-ONE

Diversions and Entertainment

busy periods in town. Farmers, with wives and children,
went in to sell produce and buy supplies. Saturday nights
outstripped the afternoons; both town dwellers and country people thronged the streets, strolling up and down
or chatting in groups. A few young men paraded with
their best girls; others stood about ogling girls who walked
in chattering little bevies. Stores remained open until midnight. Bowling alleys and other amusement centers were
well patronized; cigar stores and ice cream parlors enjoyed
a particularly lively business.

When they ate in town nearly all farmers went to
Lynch's. This was one of the biggest business places the
town could boast: grocery, bakery, and eating place, all
under one roof. Patrons of the restaurant passed through
the large front room in which were stocks of grocery
staples, tobacco, cigars, dry goods, notions, work garments, corn knives, husking gear, water jugs, horse collars,
blankets, whips, and the like. Walking through the aisle
past cases of eggs, bags of rice and potatoes, barrels of
sugar, salt, crackers, molasses, vinegar and kerosene, blocks

of rock salt, nuts in bushel baskets, displays of fruits in season, kegs of salt fish, and stacks of bar soap—including "Grandpa's Wonder" tar soap—they would note long glass cases. In these were displayed in tempting array big loaves of bread, rolls, pies, cakes, cookies, and doughnuts, all fresh from the oven.

A coffee grinder with a big hand wheel was bolted to one of the counters. Not all grocers had such machines because the majority of customers preferred to buy coffee in the bean and grind it at home. A few housewives bought green coffee beans, which they roasted in their own ovens. In most homes was a small hand grinder that was used at breakfasttime daily to grind enough for the day. "Coffee essence" was stocked by grocers and used by some as a coffee supplement or substitute. It was made from chicory root, still used in the South where many enjoy it in their coffee. Among favored brands of roasted, unground coffee in one-pound packages in stores were Lion and Arbuckle. We youngsters used what influence we had in favor of the former, because accumulations of round lion heads, cut from the bags, could be exchanged for a variety of premiums, many of which appealed strongly to children.

A door at the rear of the bakery section opened to the dining room. Just inside, at a long counter, stood a clerk to take your order for fresh, warm buns, butter, bologna, maple syrup, cookies, and coffee; the menu, always the same, offered nothing more. The clerk tore a square of brown paper from a roll and placed upon it the items you had ordered, and you paid your bill. Then, using the paper as a tray, which was to serve finally as your tablecloth, he carried your order to a maple-topped table, altogether innocent of paint or varnish, and you

242

followed, carrying your cup of coffee. The butter, a generous pat on a small plate, was of excellent quality in summer; in winter it generally had an uninviting appearance and a pronounced cheesy smell. The syrup was genuine maple and of a quality that some years later sold at several dollars per gallon. The bologna and the baked goods were top grade.

The large dining room was decorated with wallpaper on which appeared Civil War scenes in color, the figures almost life-size. Standing out prominently were big cannons with attendant caissons; everywhere were groups of the Boys in Blue, about their tents, beside their campfires, or standing at ease, leaning upon their muskets.

Women were not excluded, but I never saw any eating at Lynch's restaurant. There were several more pretentious eating places in town serving the best of food, but farmers, in their overalls and rough working garb, felt more at home at Lynch's. Probably, also, they preferred the fare because it differed so greatly from the food they ate regularly at their own tables. Accommodations were provided for about a hundred diners; the place was generally filled almost to capacity.

Our town was not big enough to attract the biggest circuses, but one or more of the smaller ones appeared every summer. Large crowds turned out to see the free parades, at that time put on as bait to lure customers to the tent shows. In one respect the parades were better than the shows in the tents; they were in the form of striking pageants, moving slowly in a line in which the spectator had time to see everything, unconfused by a multiplicity of action such as ordinarily goes on under the big top. The men and women performers in glittering costumes, the horses, the elephants, the big, gaudily painted wagons, some heavily barred, with numerous strange animals on display—about every important attraction the circus could boast—appeared in the parades. The clowns were there, all uproariously funny, always coming up with the bizarre and the unexpected. Somewhere along the line appeared the steam calliope, the sweating player seated at the keyboard back of the sizzling boiler, gloves on his hands for protection against the hot keys.

Every summer one or another of our railroads ran an "excursion" or two, advertising each long in advance. Around six in the morning of the appointed day, the long special train, made up of old wooden day coaches, pulled in at the depot. Often, the seats were jammed full of passengers picked up at previous stops. That didn't discourage the waiting excursionists at all. Tickets in hand, they good-naturedly pushed their way into vestibules and aisles and somehow found places to stand or sit. All carried boxes of fried chicken and other choice food items. All were in holiday mood, determined not only to make the trip pay the greatest possible dividends in fun and excitement but to see all of the wonders promised by the road's adman. Usually, the train got far behind its schedule—it had to be sidetracked for regular passenger trains and occasionally for some of the faster freights. Rarely did the travelers get back to the home station before midnight, several hours behind the scheduled arrival time.

All who went on an excursion were likely to have a day more trying, physically, than a day of hard work, what with the crowded coaches, heat, noise, cinders, smoke, dust, and hours of standing and tramping about; yet all enjoyed going. The trips had for us much educational value, for they introduced us (on a very limited scale, to be sure) to the outside world, and they brought us face to face with places and things we had studied about in school.

One summer a big tent was set up in town, and for a whole week the community enjoyed Chautauqua programs, one each afternoon and one each evening. The offerings, in considerable variety, were excellent on the whole. We were afforded opportunities to see and hear many notable figures—men and women prominent in the fields of politics, journalism, religion, drama, literature, and music. The Chautauqua was continued for several years, until motion pictures became strongly competitive and the automobile expanded so widely the field for entertainment and diversion.

Grange halls for housing meetings of the farmers' association or lodge, the "National Grange of the Patrons of Husbandry," were maintained at numerous points in the region. Locally, interest in the organization appeared to be at a low ebb; few of the farmers we knew were members. Farmers' alliances and co-operative associations had no known local members.

Each year a "Farmers' Institute," designed to present instruction in improved techniques of farm management and operation, was held during a period of several days in Van Wert. Meetings were well attended until a farmer in the area, regarded by his neighbors as a lazy theorist, more proficient as a smooth talker than as a practical farmer, was signed on as a lecturer to address Institute meetings well outside his own bailiwick. That seemed to reduce confidence in speakers generally, so that interest in local programs fell off.

The *Van Wert Bulletin* came twice each week to give us the news of the world, including our own little corner. Much of the time, there was also a city daily, mainly for market reports. In addition to these there were farmers' publications, religious journals, household magazines, and the *Youth's Companion*. Among books in the home library were *Uncle Tom's Cabin, Swiss Family Robinson, Robinson Crusoe, Wood's Natural History, Kidnapped, Treasure Island,* and *Twenty Thousand Leagues under the Sea.* Some of these books were given repeated readings. A special favorite was a large volume entitled *Conquering the Wilderness*. It contained stories of frontier days and related exploits of such men as Daniel Boone, Davy Crockett, Simon Kenton, Colonel Crawford, Peter Cartright, Buffalo Bill, and Kit Carson.

Thanks to the game of "Authors," which we played frequently, we learned about some of the important works of literature, coming thus to know and enjoy the writings of such masters as Dickens, Scott, Irving, Hawthorne, Stevenson, and Cooper, obtained from the public library established about 1900 and said to have been the first county library in the United States.

One day, we found about the house a Sears & Roebuck catalogue. It was only a quarter as large as catalogues issued in later years, but for us and some of our friends it was an extraordinarily interesting volume. From its pages one could gain a practical knowledge of the construction and comparative advantages of hundreds of items of merchandise (some of which, up to that time, we had known nothing about). One could study and compare prospective purchases over and over to an extent not possible in any retail store. Thus the book brought no small amount of pleasure, for anticipation is often more exciting and gratifying than actual attainment.

At an early age I was bitten by the camera bug, and mailed a dollar for a camera and a dozen dry plates, with a complete developing and finishing outfit, which I found advertised in a magazine. The camera, though nothing more than a light-tight cardboard box with simple lens and shutter, was capable of taking real pictures; the chemicals and the apparatus that came with them doubtless would have functioned as they were intended to do; but I didn't get a single picture with that outfit. Before reading the instructions, I had hastened to unwrap everything, including the dry plates. That of course ruined the light-sensitive plates, and without them no picture was possible. The outcome was very disappointing, but it did teach me one of the basic principles that every photographer must keep in mind.

Later I bought another photo outfit. This time I read all the instructions before any wrappings were removed. I loaded the camera, posed my brother and our dog, and pressed the button. Then, excitedly, I prepared chemical solutions and developed the plate in an improvised dark-room. Strange to relate, the exposure and development proved to be about right, and I turned out a picture that, if more attention had been given to the pose, would have been passably good; I had made the mistake of seating my subject with his legs extended toward the camera, so that in the picture he appeared to be nearly all feet.

The number of churches in the area was almost as great as the number of schools. Nearly all country churches were equipped with bells. In most cases they had only one entrance. Men and women might go in together, but women customarily sat at the right of the main aisle, men at the left. Usually there were two stoves, one on the women's side, one on the men's.

Two white-painted churches stood half a mile apart in our school district, one, the Methodist, the other, the Friends (commonly called the Quaker). Roughly, two-thirds of the churchgoers in the district were Methodists. The other third went to the Quaker church, about half of whose membership lived in adjoining districts. Outside the churches themselves, cleavages along religious lines were not sharply defined, although they were noticeable. Methodists at times attended services in the Quaker church and vice versa. Now and then, a preacher in one church would inject in a sermon a jab at fundamental doctrines of the other; but for the most part, these were passed over lightly by listeners and soon forgotten.

Most ministers were good men, laboring as best they knew how to promote in every way the cause of righteousness. The work they had to do was often difficult, and at times they faced hardships that, by standards of later times, were rigorous. The Methodist minister, for instance, served four rural churches. Remuneration for all tended to be pretty meager. In the parlance of the region, most of them were long-winded, sermons generally averaging a full hour in length.

In all rural Protestant churches "revival" meetings were held nightly during a midwinter period each year. Often they were called "protracted meetings," a fitting name, for they ran on night after night, sometimes three or more weeks. There was scarcely a winter week in which one could not find one of these evangelistic campaigns in progress somewhere thereabouts. In most instances attend-

248

ance was large. Most of the regular attendants were sincere believers; a few went because there was no other place to go. For some of the younger people, the meetings served mainly as social diversions; they went to meet friends and have a good time among themselves.

Both Quakers and Methodists laid down strict religious principles and rules. In consequence the community, by standards of later times, would have been considered very strait-laced. The theater, dancing, card-playing, the use of intoxicants, and laxity in observance of the Sabbath were taboo. The great majority of church members faithfully followed these teachings. Because religious leaders disapproved of them, church dinners were unknown. About once each summer month, however, one could count on a church ice-cream social not far away.

A sizable number of people in the community rarely went to church. Many of them were regular cardplayers; they worked or went hunting or fishing on Sunday without compunction; some would travel far to see a show or attend a dance, and several were pretty regular customers at saloons. Some profanity was heard, but it was generally mild and innocuous by comparison with the widespread blatant profanity of more recent times.

At Salem Methodist Church, which our family attended, Sunday school came first on regular morning programs and was well attended. Youngsters preferred it to the worship service that followed because it was much less formal and didn't last as long. If the matter had been left altogether in their hands, the majority no doubt would have ducked out and gone home before the sermon began.

No attention was given to the observance of any days on church calendars except Easter and Christmas. These were always commemorated in some way, mainly for the benefit of the children. The Christmas program was much more elaborate than the Easter observance. Often a tall spruce or pine tree was set up inside the church, nicely decorated and bedecked with lighted candles—a very serious fire hazard, what with the quantities of cotton "snow," paper wrappings and trimmings, and the inflammable needles of the tree itself. Oddly enough, no one seemed to think of this.

There was a Halloween party somewhere in our neighborhood every year. Usually, too, a New Year's party. Every two or three weeks throughout the year someone sent out invitations to a social get-together, and we all went. Boys who were "going with" girls took them; those who went stag usually found a girl to escort home.

With the exception of New Year's celebrations, at which someone got out at midnight and fired a few shotgun blasts, all of these affairs were pretty much alike. They started with some parlor games, accompanied by much giggling and, at times, boisterous laughter. Sooner or later, we engaged in "singing" games, including "Skip to My Lou," "Captain Jinks," "Old Dan Tucker," and the like. No one seemed to realize it, least of all parents who disapproved of dancing, but these were all forms of old folk dances. All sang the words, sometimes to the accompaniment of a harmonica, and we stepped in time to the music, certain words in the songs being cues to swing partners, bow to partners, promenade, dos-à-dos, circulate, change partners, and the rest. Refreshments, consisting generally of big hunks of homemade cake or pie piled high with whipped cream, were finally served. This was a signal for the breakup of the party.

Now and then, some socially minded chap who wanted some extra spending money would invite a group to an oyster supper party. After guests had assembled, his emissaries would buttonhole the boys and request cash contributions to defray costs of the entertainment. Such parties might generate a lot of fun, but the type of hospitality displayed tended to stick in the craws of male guests, especially when they considered the financial profits that accrued to the host.

Husking bees were once quite popular as social affairs in rural communities, but I never knew of any in the area about us. Our nearest approach to such a thing came after a local farmer, father of several of our classmates, died. A field of shocked corn remained unhusked. As this appeared too big a job for the boys of the family to get done before winter closed in, a dozen boys of our school went one Saturday and put on a sort of all-male husking bee. There was a great deal of pranking and horseplay, but we stuck to the task until it was done and—I report it with some pride and no small degree of wonder—done well.

Farm auctions nearly always attracted large crowds because they could be counted on to provide some entertainment and opportunities for leisurely visiting, as well as for buying livestock and farm equipment. Furthermore, there was a free lunch at noon—buns, bologna, and coffee. The coffee was served in new tin cups, that would go on the block after the repast.

Political campaigns afforded some diversion, although few took politics very seriously. Several times, we heard speeches from men who later became important figures in national affairs. I happened to be in town one day when an eloquent speaker was praising the surpassing merits of

Republican candidates who sought election that fall to offices from that of President of the United States down. Though only a small boy, I was so carried away by the words of the speaker and the music of the band that I dipped into my meager hoard of cash and invested in a long-billed cap whose front bore, in large letters, the names of the Republican aspirants for the offices of President and Vice-President of the United States. Now, my father was a Republican—quite probably that was why I also was a Republican, though an immature one. But he was by no means a rabid partisan. Apparently, therefore, he didn't care to have me advertising to all of our small world that he stood on that side of the political fence. Anyway, my prized cap mysteriously disappeared, and I saw it no more.

TWENTY-TWO

Medicines:
Amateur and Professional Doctors

SASSAFRAS WAS COMMON IN parts of the Black Swamp, but none grew in our immediate area. Many there, however, were familiar with it; they regarded sassafras tea as an invaluable tonic, and they drank quantities of it every spring. The beverage was made by pouring boiling water over roots or bark, brought in and sold in town by street vendors and grocers. All youngsters welcomed this "medicine" because they enjoyed its taste and smell.

The same cannot be said for the mixture of sulphur and cream of tartar that was administered to some in early spring. This was supposed to "purify the blood" and put one in top physical condition, after he had endured the hardships of winter. The nice thing about these two old-fashioned medicines was that, while they probably did little good, they were practically incapable of doing any harm.

Some families used a cough syrup made by steeping clean hickory bark in water, then boiling it with sugar. It was not only agreeable to the taste but it helped in relieving coughs. Mother had three favorite prescriptions for severe colds or "grippe." One was hot buttermilk with a big dash of ground ginger, well sweetened; another was hot lemonade; the third was a big onion, baked in its skin, salted, and buttered. All were pleasant to take, and all seemed to help some, especially if administered while the patient sat with his feet immersed in a basin of hot water, his legs wrapped in a heavy blanket. Deep-seated colds were often treated by rubbing over the chest a mixture of turpentine and lard, then covering the area with a flannel cloth. A "blood medicine" was made by placing sliced horseradish roots in a jug of sweet cider. This was a potion that young patients could enjoy taking. They even liked to eat the horseradish after the cider had been consumed.

Each summer some of the older women collected a supply of native medicinal plants, including boneset, pennyroyal, and catnip. Tea made from either of the latter was regarded as an excellent medicine for stomachache. Catnip-and-fennel tea was considered a sovereign medicine for colic in babies. For earache dried pennyroyal leaves were sometimes packed into the bowl of a tobacco pipe and lighted. Then someone blew through a cloth over the bowl and directed warm smoke into the affected ear. This treatment often afforded relief.

On many family medicine shelves was a glass-stoppered bottle of spirits of camphor. Fumes were inhaled when one felt faint or nauseated. The spirits were applied to affected areas for the relief of headache. Castor oil was a standard medicine. Many, including doctors, had great faith in calomel, followed by Epsom salts, as a trustworthy medicine for numerous ailments. Kerosene was used in some instances for treating frightening seizures of croup. It was swallowed by the patient, a few drops on sugar in a spoon. About every family had its bottle of "Pain Paint," which was used for numerous disorders, including "cholera morbus" and stomach cramps. Another popular medicine was "Dr. Smith's Catarrh Remedy." Whatever its value as a remedy for catarrh, it was an excellent wash for cuts and burns. Ready at hand beside these medicines, one would find a bottle of "spavin cure" or other powerful liniment for the relief of sore joints and muscles of man or beast. Often there was also a bottle of glycerin, used mainly for sore, chapped hands at corn-husking time.

Patent medicines—tonics, bitters, nerve remedies, cathartic pills, vermifuges, salves, ointments, cough drops, and headache tablets—were bought and kept at hand. At that time some drug houses were advertising and marketing "consumption cures," all totally worthless, with a cruel potential for arousing false hopes in victims of tuberculosis. Others were foisting upon gullible purchasers proprietary medicines whose fleeting, sham therapeutic effects were due solely to a high alcoholic content. Disillusioned victims of the tobacco habit could buy a drug preparation that allegedly would relieve them of all craving for the weed. Apparently, it could be effective only when taken with regular, massive doses of will power.

Two or three of our schoolmates at times wore red yarn tied about their necks; this was supposed to prevent nose-bleed. It was by no means unusual to see boys with attacks of "side-ache," induced frequently by running or fast walking, stop suddenly, lift a clod, spit where the clod had lain, then carefully replace the clod. Silly as this may appear, it was effective, presumably because the act of bending over expelled an excess of blood from internal organs. A number of pupils in school wore for weeks at a stretch little cloth bags suspended by strings about their necks. The asafetida that the bags contained was intended to ward off diseases, even smallpox. Unquestionably, this would have been effective if all the bugs responsible for disease had a sense of smell, for the stuff was quite malodorous. At times we plugged holes bored in horses' feed boxes with asafetida; it was supposed to prevent distemper. We also used pine tar for this purpose.

Every winter epidemics of such children's diseases as mumps, chickenpox, measles, and whooping cough occurred. Few parents worried much about any of them, apparently thinking that since all children would get them eventually, the sooner this occurred and brought natural immunity, the better. No one in our school had the "seven-year itch" or head lice, but both appeared among pupils of other schools. No case of smallpox was ever reported in our immediate neighborhood, but a few times there were outbreaks not far away. Then, on the advice of family doctors, nearly everybody was vaccinated.

Tuberculosis took many lives. No medicine, no treatment, was known that could stop its inroads. Diagnosing a case as "consumption" was tantamount to pronouncing a sentence of death. A few victims of the disease went to Colorado or Arizona, having been told that the climate there might have beneficial effects. The outcome was the same, whether they went West or remained at home; a steady wasting away, a decline in strength and vitality, and finally, death. Typhoid fever also caused many deaths. Cases of malaria we knew about were few. The story was altogether different during a long period after settlement of the region began. To the old-timers malaria was "ager" or "chills and fever." It was long believed to be caused by miasmatic emanations from swamps and stagnant water.

258

(The original meaning of "malaria," of Italian derivation, was "bad air.") Few of the pioneers escaped this disease, marked by chills, fever, and prostration. In some cases, it was said, even dogs became ill. Without quinine, historians of the period tell us, it is questionable whether settlers in some of the wettest areas could have hung on long enough to provide drainage and so win the battle against mosquitoes, eventually found to be carriers of the blood parasites responsible for the disease.

About 1854, outbreaks of Asiatic cholera, then sweeping the country, occurred in Ohio. The disease struck hard in the Black Swamp region, being carried in by travelers on canal packets. Populations of some canal towns were almost completely wiped out.

Fair weather or foul, doctors made their rounds, riding in buggies or sleighs behind sturdy horses that in emergencies could step off the miles pretty fast. Many of their trips must have been severe ordeals because of deeply drifted snows, ice, floods, rough, rutted roads, or seas of mud. For a long time hospital facilities were lacking; there were no clinics, no laboratory technicians, no X-rays, and few trained nurses. All babies were born in family homes. Doctors did what they could with what they had, and they made it, on the whole, a commendable job. They filled their own prescriptions from stocks of drugs on office shelves or from supplies they carried. Some of their medicines were in powder form, and each dose, carefully measured, was separately wrapped in paper.

One winter my brother suddenly fell ill. We had no telephone, but somehow we got word to our doctor. A foot of snow covered the ground, and the thermometer stood near zero. Within a short time Dr. Hines drove in. Wearing a

259

heavy buffalo-skin coat, a fur cap, huge fur gloves and arctic overshoes that buckled up to his knees, he tied and blanketed his horse. Then, big pill bag in hand, he came stamping in. He warmed his hands at the stove, then went to the patient's bedside. Unhurriedly, he studied and examined the sick boy. Outside the room, he told us it was a case of pneumonia. He wrote directions on some envelopes and placed in them medicines from his case. He gave detailed instructions for the patient's care and assured us as he left the house that he would be back at the same time the following day. The lad was really very sick. But, thanks to the care of the doctor, thanks to careful nursing by our parents, who followed instructions to the letter, he passed the crisis and within about a fortnight was recovering rapidly. This doctor was such a stickler for studying and weighing symptoms that some called him "Granny" Hines. We were glad that he was that kind of doctor. Probably no man of medicine, at that time or since, could have handled the case better, due allowance being made for the fact that some of the drugs now available are vastly more effective than any known then.

Little was done to correct defects of vision. Not only was medical knowledge of the eye and eye troubles limited, but a much larger proportion of eye abnormalities went unsuspected and undiscovered than in later times. One rarely saw spectacles worn by anyone who had not reached middle age. Generally, they were used only for reading and close work. Some bought their "specs" from itinerant peddlers who carried them, with other merchandise, in their packs; others bought them from stores in town. In either case the buyer made his selection by trying pair after pair, until lenses were found that seemed to restore good vision.

260

The small oval glasses were mounted in inexpensive steel frames. The usual cost was about twenty-five cents.

No one we knew suffered from total blindness, but the sight of some very old persons was almost destroyed by cataracts. A farmer near us was handicapped for years by eyes that were inflamed, watery, and functionally very deficient. They were believed to have been injured by poisonous dust from grain being threshed. A youth who lived near our farm was one day cutting up a yarn ball when the blade of his knife, accidentally deflected, punctured an eyeball. His doctor used leeches to clear up an infection that developed. Finally, he was fitted with a glass eye.

A small number of people lived a full span of years and retained nearly all of their teeth, sound save for the natural wear of use. The less fortunate majority, however, beginning early in their lives, suffered from tooth troubles. Their natural teeth, sometimes patched and repaired, often with wide gaps between them, might serve until middle age; then they had to resort to "store" teeth. Some never visited a dentist except to have bad teeth extracted and to be fitted with artificial grinders. Others had their teeth inspected and cared for at regular intervals. Dentrifices and tooth brushes were in much less common use than in later times. A man who lived well past the age of seventy, retaining all of his teeth in sound condition, attributed their preservation to a daily scrubbing with soap and water on a cloth.

Occupational hazards have always existed for the farmer. They have increased in number as the use of machinery has increased. Early manufacturers of farm machines made them reasonably safe and foolproof; later manufacturers have given even greater thought and effort to assure those qualities. Yet, through the years, farm machines have caused a great many casualties. Carelessness or ignorant operation can be blamed in many cases; too often built-in safety devices are ignored.

A young man we knew was mowing in a hayfield when a heavy wad of green grass became entangled on the cutterbar of the mower. As he worked to remove it, the horses moved forward a step or two. The knives caught his arm and mangled it so badly that he was crippled for life. This was not the fault of the machine; there was a clutch in the gearing that drove the cutting blades, but he had neglected to disengage it. At a later period another man was driving a tractor when a leg of his overalls caught in the rotating shaft of the power take-off. Instantly, most of the clothing was ripped from his body and wrapped on the shaft. In all probability, if the garments he wore had not been old enough to tear easily, he would have been whirled and beaten to death. Here, again, negligence was responsible —he had failed to cover the shaft with the safety guard provided by the manufacturer. A thresher lost an arm at the elbow when it was caught in the gears of a steam engine he was operating. Reasonable precaution could have prevented this accident.

Boilers of threshing engines could blow up. No little danger attended the use of the threshing machine. Arms, hands, or clothing might be caught in belts, cylinder, or other moving parts. Shredders, used for husking corn after it had been cut and shocked, occasionally maimed or severed hands or arms. Mechanical pickers, which husk ears from standing corn, are fully as dangerous.

Serious trouble could result if a mower or other machine tore into a nest of bumblebees. The insects, bent on revenge, would attack furiously. If the horses were stung, great presence of mind and skillful maneuvering on the part of the driver—often also being stung—were required to avert a runaway that could be disastrous to man, horses, and machine. The only safe procedure was to keep a tight grip on the reins and hustle the whole outfit from the scene as fast as possible.

Working in timber may involve considerable risk to life and limb. A shifting wind or a miscalculation in the use of saw and wedges in felling a tree may cause it to go down prematurely or in a direction different from the one planned. In falling it may break off branches and hurl them with great force. In his youth my father was one day chopping in the woods when his ax caught in an overhanging branch. This deflected the blow and the sharp bit split open his kneecap. Blood poisoning that ensued nearly cost him his life. After the wound finally healed, the joint remained stiff and painful, a permanent handicap.

Farmers commonly rubbed the budding horns of calves with a dampened stick of caustic potash and thus prevented the development of horns. Mature cows whose horns had been allowed to grow were not dangerous to humans, but if they were too prone to hook other animals, it was a common practice to saw off the horns. Bulls were often allowed to grow horns because the appendages helped to establish the breeding of the animals, but horns greatly increased the hazards of working around them. With or without horns, bulls tend sooner or later to become belligerent. To be on the safe side, we inserted a brass ring through a hole cut in the lower septum of the animal's nose and fastened it there permanently. This is such a tender spot that it is easy to control the most unruly bull by keeping a good grip on the ring.

We once sold a young bull to a neighbor, warning him that the animal, not yet adorned by a nose ring, was becoming cross and would bear close watching. Brandishing a length of heavy fork handle, he assured us that he would be able to manage. After being driven a few hundred yards the bull, planting his feet firmly, assumed a very menacing attitude. Up went the stick and with a resounding whack it was laid squarely across the bovine forehead. The bull fell as if struck by lighting. A minute or two later, blinking rather foolishly, he staggered to his feet and allowed himself to be driven away, meekness personified.

Mature boars grow long, sharp tusks as wicked and formidable as those of their wild forebears. This makes them a serious menace to other animals as well as to humans. To disarm a long-tusked animal, we maneuvered him into a ruggedly built pen and lured him with corn to a point where we could slip a strong lasso-like rope over his upper jaw. Then, closing the noose and drawing him close to a stout post, we knocked off the vicious teeth with a hammer.

The tip of a hog's long, pointed snout is as tough and wear-resistant as rubber. It is ideally designed and built for digging and plowing into the ground. The equipment of course is an inheritance from wild ancestors that lived largely by foraging under the surface of the ground for grubs and vegetable growths. It would be ruinous to fields to allow hogs to root at will—given even a slight incentive, an ambitious porker can burrow deeply over a large area in an incredibly short time. So the farmer makes sure that all of the root is taken out permanently early in the pig's career. This is done by inserting and closing securely in place a small steel ring right through the tough cartilage that forms the cutting edge of the porcine snout. The ring thereafter spoils all the pleasure and fun of rooting.

264

The hog is an extremely stubborn and suspicious animal. Though he will not always fight to defend himself, he will readily take up the cudgels in defense of another individual of his kind, even though the two may be anything but friends. If we caught a pig for any reason and it began to squeal, or if one somehow accidentally trapped sent up a raucous protest, we had to keep a wary eye on all others of the tribe because, though we might be humanely aiding the squealer, there was real danger that part or all of them would attack us.

Buck sheep often become cross, but their victims are likely to suffer more from upset dignity and a sudden burst of anger than from physical injury. They attack by making a swift run and butting an unsuspecting human, preferably from the rear. They are hesitant about attacking a grown man, but a boy who goes near one will do well to keep a sharp eye on him. A buck we had once advanced toward me with pronounced hostility. Seizing a broom, I poised it at the ready above my head. Taking that as the challenge he hoped for, he dashed at me full tilt like a white charger. I swatted forcefully; but before the blow landed, his head struck my midriff with resounding impact, and I went down for the count.

Rattlesnakes and moccasins were occasionally reported in the region, but we never saw any. Garter snakes were common. There were some water snakes, a few spotted "house" snakes (often called milk snakes because of a popular notion that they sucked milk from cows), and a few blue racers, the latter mainly in fields and woods. These reptiles were non-poisonous, all actual allies of the farmer because they destroyed many rodents. Cornered, some might bite, but they were incapable of doing real harm to man. Nevertheless, most people feared them as much as if they had been venomous, and killed them on sight. One of the boys at school became expert at catching garter snakes in his hands. At times he amused himself by using them to frighten girls. Finally, by a quick jerk similar to the movement in cracking a whip, he would snap off their heads.

Here and there poison ivy grew over fences or on trees. Some fortunates, with natural immunity, could work about it and handle it in perfect safety. Others were so susceptible to the poison that merely walking near a plant would cause their skin to break out in a rash accompanied by intense itching and burning, which induced great discomfort. Many suffered from accidental contact with some part of the plant, living or dead. In one instance a case of blood poisoning developed when a tormented victim scratched open an inflamed patch of skin.

Dynamite, used extensively for blasting out stumps, was treated with such respect that no accidents occurred. Recalling the toy cannons that some of us made and used on the Fourth of July, I am forced to the conclusion that a merciful Providence must somehow have intervened to save us from maiming or destruction. We made them from pieces of gas pipe or from long, large-caliber rifle shells anchored to blocks of wood. Loading them with black powder, we set them off with matches. Fireworks were cheap and plentiful. Quantities were used, but no one was hurt, and no fire damage occurred.

266

TWENTY-THREE

Superstitions and Tales

NO ONE PAID MUCH ATTENTION
to the black cat superstition, but some were convinced that
breaking a mirror or walking under a ladder could not fail
to bring bad luck. A few were afraid of the numeral 13 and
regarded the thirteenth of a month that fell on a Friday as
particularly ominous. Giving a knife or other edged instrument as a present was disapproved of by some because of
the danger that it would "cut friendship." Some believed
that if one cut his nails on Sunday "the devil would be with
him all week"; also, that if human hair was used by birds
in constructing a nest, the owner of the hair would suffer
from headaches. Girls were seriously advised against
whistling because "a whistling maid and a crowing hen
never come to a good end." Finding a pin portended good
luck, particularly if the point happened to be toward the
finder. If one had a wart, he was advised to sell it for something of value—a pin or a penny; then, if he hid the token
of sale where it would never be found, the wart would disappear. In some cases it did actually fade out completely.
Probably few parties to such transactions would have believed that this happens fairly often, even when the
blemish has not been "sold."

A few went to great lengths to get their garden and field crops planted in the "right sign" of the moon. If the weather or other condition prevented this, they reluctantly planted at another time, certain that results would be unsatisfactory in some important respect. They insisted that it was the height of folly to butcher pigs unless the moon was in the proper phase. If the "sign" was not right, they declared, the fat would "fry away." Some were certain that early garden stuff, to do well, must be planted on Good Friday. No matter how unfavorable the weather or how wet and clammy the ground, they strove mightily to get their gardens started that day.

Two "witnesses" told in all seriousness of having seen about his late home the ghost of a man who had taken his own life. Other accounts, rather hazy as to time and place, were given about a man who became so wicked that the devil came one night to claim him as his own. The intended victim took to his heels and ran so fast that he finally left the panting Old Nick far behind. There was no little curiosity as to whether the chap was scared into abandoning his evil ways or the devil persisted until at last he nabbed him.

A young friend and I, both about twelve, went coon-hunting one very dark night. We had no hound, but we carried guns and a lantern. Doubtless influenced by the darkness and the novelty of the adventure, we talked of ghostly occurrences that we had heard of and dwelt at considerable length upon various weird and uncanny matters. When we got well into the woods, we somehow felt the presence of a deadly enemy, maybe lurking behind a tree or a stump. Shadows cast by the lantern's beams took on strange forms, unearthly and menacing; a breaking twig sounded exactly like the cocking of a gun; cries of night birds made us think of "catamounts" and bears. The sighing of the wind, the rustling of leaves, the darkness, all assumed unwonted qualities, mysterious and ghostly. We didn't see a coon or any other wild creature. As a matter of fact, our state of mind became such that we didn't want to. We had gone only a mile or two into the woods when—our hearts pounding, our hair bristling, and our nerves tense—we decided that we had had quite enough of coon-hunting. We turned about and started homeward, walking very fast and clutching our guns tightly, all the while glancing stealthily right and left over our shoulders.

Few of the people about us actually gave real credence to the supernatural; however, nearly all had an unflagging interest in everyday natural happenings, particularly if they were a little unusual or bizarre. Practically everyone had ears open for accounts of amusing or exciting goings-on. They enjoyed hearing and passing along anecdotes, especially any that might have a sparkle of humor. Some of the tales that were told and some of the incidents that aroused interest or brought laughs are recounted in the paragraphs that follow.

Our neighborhood was set agog one midsummer day by reports that a "wild man" was at large, wandering through the tall corn over the countryside. It developed that the wanderer was a poor old fellow suffering temporarily from strange delusions incident to a senile disorder. About the same time we were told that somewhere "up north," "White Caps" were about to close in on a man alleged to be a wife-beater. There was an overtone of mystery here; up to that time, no one had ever heard of White Caps. I have never heard of them since. It seems probable that the name had been adopted by some local self-constituted vigilante-type group, perhaps a sort of small-scale Ku Klux Klan.

A wide local area was electrified one morning by grapevine reports that a horse thief from another county had passed through town and was heading north, our sheriff pressing closely behind him. In the fracas that followed, when contact was made, the lawman shot and killed the fellow. Half of the people in the county visited the undertaker's room the next day for a look at the dead man. Many felt that the sheriff, a brave man who, in his time, rounded up many a lawbreaker, possibly was a little too quick on the trigger in this instance.

Not long after that, everyone was shocked by a shooting tragedy at the home of a widely known farmer. Incensed by a series of petty thefts from his premises, he had begun a nightly vigil, determined to catch the thief. As he watched one evening, a figure dimly outlined in the darkness appeared and moved directly toward the barn door. He blazed away with his shotgun, and the supposed intruder fell lifeless to the ground. The family was horrified to find that the victim was a young nephew of the man's wife who had walked over from his farm home for an overnight visit. The law ruled that the shooting was an accident. Public opinion held the farmer guilty only of ill-advised, overly-hasty action and all in the community, though grieving for the death of the youth, were more disposed to sympathize with the farmer than to censure him. The man himself was inconsolable and never ceased his self-reproach.

An aged man with his wife and several grown sons came to live on a little farm not far from ours. The father, kindly, neighborly, and industrious, won the respect of everyone. The boys, however—big, strong fellows, always well dressed—were never known to do any work. Their lack of any visible means of support aroused widespread suspicion and was the subject of much discussion.

After a few months there were reports of burglaries in surrounding areas. The finger of suspicion pointed directly to these young men, but officers were unable to find any ironclad evidence against them. On Christmas Eve the home of the Sunday school superintendent, half a mile from the home of the brothers, was burglarized. No one doubted that this was their handiwork, because they knew the lay of the land and could be certain that the superintendent and all members of his family would be at the church that night for the Christmas program. As tangible evidence was lacking, no action could be taken against them. Not long afterward, however, the farm was sold and the family moved away; evidently, neighborhood suspicion and resentment had become too strong.

Several generations of fun-loving pranksters had depended for a big share of their laughs upon the snipe hunt, invented far back in backwoods days. Apparently unaware that the ancient gag had become superannuated, a group singled out a young chap as a gullible tenderfoot and invited him to go with them to hunt snipes. With a well-feigned lack of sophistication, he accepted. When they had penetrated deeply into a thick woods, they handed him a bag, directed him to stand very quietly and hold the bag wide open while they worked through the brush and beat out the snipes. The birds, they assured him, would zoom in and dive straight into the bag.

He took his stance and held the bag exactly as directed.

Immediately after they were out of sight and hearing, he stole quietly from the spot and ran swiftly over a wide detour. He met them as they emerged from the woods, slapping each other smugly on the back and laughing uproariously. He never enjoyed anything so much in his life, he declared, as seeing the sheepish looks that suddenly came over the faces of the other "hunters."

A farmer told of going with a binder fresh from the factory to harvest a field of rye for a man on a neighboring farm. Noting the exceptionally heavy, badly tangled straw, he expressed concern that the machine might be damaged, if not completely ruined, trying to handle that crop. "Why worry," replied the neighbor, "the thing's quarantined, ain't it?"

A big fellow with oxlike strength, able to do with apparent ease the work of two average men, often helped us with farm work. He was heard to complain, as he spaded in a long trench for a tile drain, that the work became a little "monopolous" at times. Another chap stood several minutes watching a hog as it wallowed in a filthy mud hole. "Well, sir," he observed, "they sure named them there animals right when they called 'em hogs."

A jovial Irishman with an ever-present sense of humor and an oft-expressed musical preference for "something quick and devilish" lived on a farm near ours. "I've noticed," he said one time, apropos of March weather, "that, so far, when I've lived through March, I've lived through the whole year."

He had a story about a farming community that was suffering acutely from lack of rain. Crops were withering

274

and parching in the fields; wells and cisterns were going dry; farm animals were on the point of perishing. Desolation was everywhere. Finally, it was decided to get all of the people together and pray for rain. An old fellow whose crops had been hit very hard led off with, "Lord, send us rain. We don't want no cloudburst. No jimmycane. We 'druther not have no lightnin'. No thunder, neither. Just send us a nice drizzle-drazzle that'll last two or three days."

A painfully pious old chap in town was daily and hourly distressed by the sin and depravity that he saw in the world about him. He was working one day with a gang of men digging a deep trench along a prominent street when a circus was in town. As the spectacular parade approached, all of his fellow workers clambered out of the ditch to watch it. Old Mac, however, stretched himself face down in the mud at the bottom, so that his eyes would not be offended by the sinful goings on. Years later, Mac became proprietor of a grocery store. A customer one day asked for some tobacco. "You can't buy none of that there wicked stuff in here," he declared. "If you're bound to have it, go across the street; they sell it over there, at the devil's store."

A farmer kept a flock of hens that laid exceptionally well. When his son Chet, a confirmed practical joker, suggested that they might do even better if fed more tankage, a generous amount of this rich feed was added to their ration. The following morning Chet stealthily returned to the nests a dozen of the eggs collected the previous day. That evening the older man stared in wonder at the immediate "results." Chet next morning slipped two dozen eggs back into the nests. Results this time were phenomenal—in just two days the flock's "production" had been brought very close to 100 per cent! Chet held the peak output a few days, then brought it back gradually to an honest normal.

A quaint old character was widely known for his big repertory of tall stories. He told them all as sober truth, unaware or unmindful of discrepancies or contradictions. Here is a typical yarn attributed to him:

"One day when I was a young feller, I had a keen hankerin' fer wild turkey. I loaded up my gun and walked out acrost the medder to a field of wheat. Just ready fer the cradle, that there wheat was. Big heads, tall, straight straw. As I was a-walkin' along the fence, I seen a big gobbler's head a-stickin' up above that there wheat. Well, sir, I took aim, and I bored that there turkey right through the head. I tied his legs together, and I slung him over the barrel of my rifle. That there bird was so big that as I walked along, the gun over my shoulder, his head drug in the snow."

Another yarn-spinner was given to telling what might fittingly be called transparent lies. All were simple tales that could not possibly do any harm. Although all bore on their faces the indelible imprint of falsehoods, he told them as solemn truth and evidently expected his hearers to believe them. One of these tales ran to the effect that when he was a small boy, playing with his foster mother's spectacles, he accidentally broke a lens. "I was sure scairt," he said, "for I knowed that Mother would whale me good if she found out. All at once I remembered that a lot of glass was scattered over the ground where a winder had been broke back of the house. Well, sir, I went out there, and I hunted and scratched around till I found a round piece of glass. I pushed it into the frame, and it fit just like the one that got broke. I snuck in and put them specs back on the shelf where they belonged. Mother wore 'em a long time after that and she never knowed the difference."

A man whose veracity was everywhere regarded as above reproach came up one time with a story that doubtless could have made him a respected member of the Liars' Club. "For a long time," he said, "I had a lot of trouble with hot feet. Often, when I retired at night, they seemed fairly to sizzle. To be at all comfortable then, I had to place them in a pail of cold water on the floor beside the bed. This brought such comfort that nearly always I got a good sleep. But the arrangement had its drawbacks, especially when the weather got cold. One winter morning, for instance, when I awoke, my feet felt just right; but I had some trouble getting them out of the solid ice in that pail."

An odd, bewhiskered old fellow was in the habit of wandering more or less aimlessly over the countryside on foot. We were told that he particularly liked to come upon a farm where men were threshing. Just as all was in readiness for the customary big dinner, he would walk to the house, knock the ashes from his smelly old pipe, and maneuver himself into the dining room. Looking over the table, he would say to the women, "I believe I'll just take two or three pieces of that chicken, a few slices of that ham, a few rolls, a wedge of that pie, a coupla bananas, some of them pickles, and a hunk or two of cake. Then I won't stay for dinner."

A minister who was fond of hunting went one day with another Nimrod from whom he heard this account: "I was a-huntin' in the old oil field over in the Evans neighborhood a coupla years back. All at onct it got dark, purt-near as black as night. I knowed a storm was a-gonna cut loose any minute so I run to an old boiler a-layin' there in the field. I no more'n got into the firebox and pulled the door to behind me than she started. The wind screeched and howled, a-blowin' a sockdollager shindy. Made that there old boiler creak and sorta tremble like, as if it was about to start a-rollin'. Rain poured down 'bout like somebody up in the sky was a-upsettin' a big lake. And the lightnin'—man, as I peeked out, it streaked and flashed everywhere. Like t'blind a feller. Reminded me of a great mess of big, red-hot snake tongues. Thunder rolled and rumbled. Sounded so loud in that there firebox I thought it was a-gonna make me deef fer sure. Never seen no such storm in all my born days."

"Were you scared?" asked the preacher.

"Was I scairt! Man, I squatted in that there sooty old firebox, a-shakin' like a leaf and a-prayin' like hell."

A man in town had been talked into taking out a policy of a few thousand dollars in a mutual life insurance company. For a long time, assessments remained reasonably low, and he had no complaints. But, as policyholders began to die off, costs to survivors increased steadily and at length became burdensome. One day the collector called for a payment, the heaviest so far asked for. The old

278

fellow dug up the money. As he handed it over reluctantly, he reviled himself for starting with the insurance. "Costs too much," he said. "More all the time. And, by Cripes, I'll bet a dollar it will be just my rotten luck to be the very last galoot on the list to die."

If someone claimed for himself credit for an accomplishment and it appeared that said credit might rightfully belong to another, the braggart was pretty sure to hear the retort: "Yes, Betsey and me killed the bear." This had reference to an old tale of frontier days. According to the story, a big bear invaded a settler's cabin. The frightened husband beat a swift retreat; from his hiding place under the bed, he poured forth words of encouragement and advice to his wife who, armed with a big poker, valiantly stood her ground against the intruder. Finally, the good woman, with a lusty blow, laid Old Bruin low. Thereafter, the husband took great pleasure in relating to all who would listen how "Betsey and me, we killed the b'ar."

In case someone, working at a tough, wearing job, showed symptoms of discouragement, his friends, urging him on, might remind him that it was "the makin' of the pup." This harked back to an ancient tale of the region about a man engaged, with his son, at training their young dog to fight. On all fours, he slugged the pup, baring his teeth and growling realistically. The husky pup, entering heartily into the spirit of the lesson, vigorously chewed and mauled his instructor. After a few minutes, the father, scratched and bleeding, was on the point of abandoning the project when the son, from a safe spot on the side-lines, yelled: "Don't give up, Pap. Don't give up—it's the makin' of the pup!"

A man in the community and his two stalwart sons all seemed to recognize in a special way the uncertainty of human life, and even to entertain some doubt as to the permanence of our planet. In proposing or promising to do something at some future date, each nearly always added the qualifying proviso: "That is, I will if I live and the world stands."

After he started working by the month for a new employer, a local farmhand, impressed by the fact that his boss belonged to the Improved Order of Red Men, lost no time in becoming a member. After that, he always referred to himself and the other man as "Red Brothers." This appealed to the sense of humor of neighbors who, observing the two at work, would always remark, "Well, I see the red brothers are hauling manure"—or "slopping the hogs," as the case might be.

TWENTY-FOUR

Far-reaching Changes

WHEN I FIRST BEGAN TO TAKE
note of farm activity about me, in the 1890's, some of the
tools and implements that farmers worked with were little
different from those that farmers for generations before
them had used. They had a number of items of field
equipment that served passably well. Some of these, including plank drags (or "clod-crushers"), land rollers made
from log sections, corn row markers, and mudboats were
homemade; several were the handiwork of local black-
smiths; the remainder were factory products. The use of
a few of those horse-drawn contrivances involved much
hard work on the part of the operator, not to mention
the many miles of walking. Farmers today would scoff
at most of them as crude, clumsy, and inadequate.

There were drills for sowing seed grains, but where
these machines could not well be used, it was not unusual
to see wheat, oats, rye, or grass seed broadcast by hand.
This method of distributing seed probably differed little
from that of the sower who "went forth to sow," as in
the biblical parable. (By analogy, the word "broadcasting"
in later times has come to refer to the transmission of radio

and television programs over wide areas by means of electromagnetic waves.) The sower carried the seed in a bag or other receptacle within convenient reach. He swept his arm in a wide horizontal arc before him as he walked, releasing the seed held in his hand gradually, so that it was distributed evenly over the ground. His reach into the receptacle and the swing of his arm were in rhythm with his steps. He set stakes to guide himself in a straight path and to regulate the width of the area covered, measuring distances by pacing. Allowance had to be made for wind force and direction. The field was finally harrowed to cover the seed.

I saw men cutting grain with cradles, bunching the straw with wooden rakes and binding it into sheaves by hand—the ground was so stumpy that no machine could be used. On a nearby farm was a "self-rake" reaper, which had seen much service in cutting standing grain and depositing the straw on the ground in orderly bunches for laborious hand-binding into sheaves with twisted wisps of straw. This machine was then used on that farm only occasionally, to cut and bunch ripe clover to be dried and hulled for seed. As a grain harvester, it was obsolete; the self-binder, as it was called at first, had some time before taken over the grain-harvesting function.

At about the time I viewed the harvesting scene, I saw a portable steam engine that had to be moved from place to place by horses. It had superseded the treadmill and horse-power mechanisms that utilized the muscular power of horses for driving threshers and other machines. This engine, too, was then outmoded, having been supplanted by the steam traction engine, which could be moved at will under its own power.

The coming of that traction engine could be regarded

as marking the inception in that area of farm mechanization, to be characterized as time went on by the general substitution of mechanical power for the muscular power of men and animals. After that engine, new farm machines and improvements in old ones began to come at a somewhat accelerated pace, one development often leading up to, and paving the way for, another. Although all of us there were witnessing the local beginning of an era in which would come a profound revolution in farming operations and rural life, the development had been so gradual that there was no open-eyed wonder. We were all too close to see in true perspective, to evaluate correctly what was happening, or to foresee what would follow.

Within a comparatively short time the steam engine had to give place to the far more versatile and useful tractor, powered by an internal-combustion engine. It could perform the steam engine's sole task of driving machines by means of a belt; in addition, it was designed for the operation of farm equipment in fields. Today powerful tractors that handle with ease big gang plows and other equipment for rapid and efficient tillage are commonplace. Tractor-powered machines reduce sharply the time required for sowing, planting, cultivating, and harvesting. Thus it is possible for one man to handle well five or six times the acreage that one man could care for in premechanization days.

Rivaling the tractor and concomitant equipment as boons to farmers of the Black Swamp and elsewhere is electricity from central generating stations. Perhaps the most important of the contributions it makes is the automatic pump, which provides a supply of running water and so makes available to the farm home a sanitary plumbing system. Going further, to provide for rural people the comforts and advantages enjoyed by urban dwellers, electricity also makes possible on the farm good lighting, an impressive assemblage of devices and appliances that save time and labor, efficient central heating, air conditioning, the mechanical refrigerator, and the deep-freeze unit. It widens farm families' horizons of culture and entertainment by bringing to them the best in radio, recorded music, and television—also, it must be admitted regretfully, much that is exceedingly poor.

Coming years before electric service, macadam roads, which took the farmer out of the mud and put him into closer touch with markets and the outside world, were improvements of inestimable value. But those roads were not good enough for the automobile; soon after this vehicle came into extensive use, it stimulated such insistent demands for their betterment that, within a surprisingly short time, they were extended and made more serviceable. Now, the rutted one-lane pikes, dusty when dry and sloppy when wet, have become dustless, hard-surfaced roads that are integral parts of a vast nationwide network of highways.

It would be difficult, if not impossible, to appraise fully the influence and the impact of the automobile upon American life, rural and urban. Perhaps there could be no better summarization than that made years ago by President Hoover's Committee on Social Trends. It reported, regarding this machine: "It is probable that no invention of such far-reaching importance was ever diffused with such rapidity, or so quickly exerted influences that ramified through the national culture, transforming even habits of thought and language."

Aided by the automobile and good roads, the telephone brings the best of medical care and hospital facilities

284

within easy reach of the farm family; most farm babies are now hospital-born. Being available for use in emergencies, the telephone materially enhances the security of all ruralites. From the first it has been of great service in facilitating the transaction of business and in linking farm families more closely with neighbors and friends.

Several crops for which no market formerly existed are now grown extensively on Black Swamp farms, widening the field of productive activity and augmenting incomes. Hybrid corn, bred to assure a number of desirable qualities formerly lacking, is increasing yields and putting an end to some of the problems that worried growers of earlier times. Seeds of other grains have been materially improved. Careful selection and breeding have produced better cattle, sheep, hogs, and poultry.

Thanks to these advances, to continuing assistance from agricultural science, and to modern machinery that multiplies individual productive capacity, the average farmer enjoys a better income than his predecessors had. He lives more comfortably; he can find time for recreation, for travel, or for vacations if he wishes. He may even sojourn occasionally in a sunny clime to escape the unpleasant features of a northern winter. Let no one assume, however, that his life is one of carefree ease. He has a great deal of hard work to do. Furthermore, he is beset by disquieting worries peculiar to the farming business. The more serious of these worries stem fundamentally from the long-standing problem of crop surpluses and the resulting depressed prices of agricultural products. He is disturbed by the disparity between prices he receives for commodities and prices he must pay. He finds farm machinery generally so costly that owning a complete outfit may require a large investment—variously estimated at from $30 to $80 and more per acre for the larger farms. To make matters worse, a large proportion of this equipment is needed and used only during a short period each year.. The remainder of the time it stands idle, earning nothing, while interest charges run on.

Before the advent of powered equipment, we and others on farms in the region practiced diversified farming. That is, we kept domestic animals in numbers economically suited to our acreage, and we grew a variety of crops, most of which were fed to those animals. We counted on the sale of mature livestock and surplus grains for a cash income. With proper management under this system, little, if any, of the soil's nutritive elements need be permanently lost from a farm.

For a long time straw, cornstalks, etc., were burned on a few farms as unwanted refuse. The more enlightened farmers considered this a gross waste of potential humus. They took pains to return these and as much as possible of other organic materials to the land. They followed a sound program of crop rotation, and they grew clovers, knowing and valuing the peculiar ability of such plants to enrich the soil. By these means they maintained the fertility of their acres with no noticeable impairment, unaware of any need for the chemical fertilizers now used on Black Swamp farms in an amount that totals thousands of tons yearly.

With mechanization has come a trend away from diversification toward specialization, the latter often somewhat limited or modified. In addition to the establishments where more or less diversified farming is still practiced, there are a number of farms that have no livestock of any kind—not even chickens. In many cases operated in conjunction with neighboring acreage, these farms may produce only grains or other marketable field crops. Here, of course, commercial fertilizers are indispensable. Some of the farm specialists concentrate on dairy production, others on poultry and egg production, or on growing potatoes or fruits.

In some instances two or more farms have been consolidated under one ownership. A new name may appear upon the barns, unless they and other buildings are moved away, as they often are, to reduce valuation for tax purposes. Thus we see farms increasing in size and diminishing in number. For the most part, however, they remain family enterprises.

A number of owners of sizable farms have adjusted to changed conditions brought about by mechanization by working at full-time jobs in town (sometimes thirty or more miles distant) and operating their farms as sidelines. The time was when retired farmers in numbers moved to town; now, as a rule, they choose to remain in their comfortable farm homes. They rent their fields to neighboring farmers, as do some of the younger men about them who live on their farms but work at jobs in industrial establishments. There has been a noticeable migration of farm people to urban or suburban areas and industrial employment. Concurrently with this displacement of country populations has come an impressive movement of people and industries into suburban and rural areas. Some farms that lie relatively near the larger towns and cities are owned by business and professional men. Regardless of size, they are commonly called "estates." Rather often, the owner lives in a good house on the property, driving to and from his work in town. Generally, a less pretentious house is provided for the man who operates and cares for the establishment.

The status of barns has been drastically affected by the elimination of livestock that has occurred, together with the use of combines, field balers, and other late-model machinery for harvesting and handling corn crops. Not infrequently, one now sees large barns, originally intended to shelter livestock with ample supplies of feed, standing virtually empty and unused, deteriorating rapidly in some cases because of lack of care. New barns that are put up tend to be rather small; construction materials in them comes almost exclusively from commercial lumber yards.

Like the big barn, the silo, formerly widely used for storing succulent feed materials, has become a useless adjunct on a large proportion of farms. Neglected and forgotten, some are gradually falling into ruin. Windmills have been generally phased out, supplanted by electric motors for pumping water. Towers of some now support television antennas; scores of others, battered and rusted, stand as creaking, ugly eyesores.

The faithful horse inevitably became a casualty when farmers generally gave in to the lure of the tractor. To all who are fond of the animals, it is a matter of deep regret that, with the exception of a few riding horses here and there, horses are practically extinct in the region. Occasionally, one comes up for sale at an auction. Animals that don't qualify as saddle horses are sold at a few dollars a head as "foxers," to be slaughtered and used as food for stock on fox or mink farms or for carnivores in zoos.

288

Most of the old one-room school buildings have gone by the board. A few have been left to the ravages of time and the elements. Occasionally, one is found in use for farm storage of some kind. A number have been remodeled and adapted to use as family dwellings. Numerous church edifices, also long familiar landmarks, are no more. Two in the home community, including Salem Methodist, were demolished by tornadoes a few decades ago. The Salem congregation, jointly with congregations from three other churches—all much smaller than in former times—erected a new building that all, as a unit, now use. The automobile and improved roads have furthered the consolidation of a number of other church organizations.

A goodly number of older farm houses have been remodeled, becoming more comfortable, more convenient, and more attractive. New houses generally embody the best in modern design and arrangement. It is evident that more farm families are giving thoughtful consideration to the architecture and placing of buildings and to the landscaping of adjacent grounds. Unfortunately, however, many of these people, like others in the region, generally neglect field growths of noxious, unsightly weeds, such as Canada thistle, wild carrot, and Jimson weed, allowing them to flourish and spread freely.

Wooded land is diminishing in area. Most remaining stands of timber are thin and show symptoms of dying out. Everywhere, the Dutch elm disease has made devastating inroads. Rarely, it seems, is any effort made toward preservation or renewal of timbered tracts. Few bother even to remove unsightly dead trees. It is gratifying to note, however, that natural agencies have continued active in propagating trees and shrubs, so that the countryside as a whole retains its appeal to lovers of natural beauty. Here and there at well-chosen sites—in numerous instances at points of historical significance—public parks are maintained for the pleasure and convenience of local people and travelers. Probably no taxpayers' money buys more in sound value than that spent for establishing and maintaining these parks.

Native wild animals generally seem to have adjusted so well to the loss of natural, thickly wooded habitats that some species are seen in greater numbers than formerly. Groundhogs, for instance, once somewhat rare, now appear to be the most numerous of all wild creatures, strongly entrenched in old fencerows and along wooded banks of numerous open streams. Occasionally, they become destructive pests, causing serious damage to field and garden crops.

In any study of Black Swamp history one of the most striking of the emerging facts would be the paradox that a region that only a few generations ago was plagued by an overwhelming perennial surfeit of water is now so dehydrated that during a considerable part of the average year large sections suffer from a critical shortage of water. This has become a serious matter that is steadily worsening as populations grow, as household demand for water increases, and as water-using industries expand. Of late, serious consideration has been given to proposals for the installation of large-capacity pipe lines to carry water inland from Lake Erie.

TWENTY-FIVE

Recent Looks at the Valley

IF HE IS UNFAMILIAR WITH
local history, one who visits the Maumee Valley may
wonder at the numerous reminders of General Anthony
Wayne that appear—monuments and memorial parks, dedi-
cated in his honor, as well as highways, streets, schools,
and an important city that bear his name. The fact is
that the admiration and respect long felt in this region
for General Wayne parallel rather closely the high esteem
in which General Robert E. Lee is held in Virginia. This
dates back to 1792, when President Washington, remembering the general's demonstrated soldierly qualities—he
was widely known as "Mad Anthony" because of his
dashing bravery in numerous Revolutionary War engage-
ments—sent him into the Ohio country as commander in
chief of the Western army. His assignment was to turn
the tide of British-Indian gains against American arms.

The first significant achievement in the decisive success
of this mission was his defeat in 1794 of the Indians at
Fallen Timbers, just off the Maumee River, between the
present towns of Waterville and Maumee. The second,
largely the result of the first, was his negotiation the
following year of the Treaty of Greenville, which opened
the Northwest Territory to white settlement. Not only

did his successes here enhance his reputation as an able military leader but they did much to validate the claim of the valley to historic fame.

We are told that as the general, with his command, advanced northward down the Auglaize River toward its juncture with the Maumee, the Indians, in a large cluster of villages on the shores of both streams, having been warned by their spies, fled precipitantly—they had come to regard Wayne as a formidable antagonist who, they said, never slept. They left behind apple and peach orchards and extensive plantations of corn and squash. The soldiers laid waste the fields and later destroyed corn and other crops growing farther down the Maumee. This brought the red men to the verge of starvation the following winter and, together with the trouncing they got at Fallen Timbers, made them willing to sit about the Greenville council fires and smoke the pipe of peace.

Surveying the fort that his men had built at the mouth of the Auglaize, Wayne exclaimed: "I defy the English, the Indians, and all the devils in hell to take it!" Thus it became Fort Defiance, whose site, with restored earthworks, is now preserved as a memorial. The settlement that grew up about the fort, now a thriving city, was named Defiance, as was the county of which it is the seat. Some twenty miles beyond the Ohio-Indiana line, where the St. Marys and St. Joseph rivers join to form the Maumee (called "Three Rivers" by the Indians), General Wayne built a stronghold that soon got the name Fort Wayne. The hustling Hoosier city now bearing that name surrounds the site of the fort. On the Wabash River in Mercer County stands the town Fort Recovery, named for the fort that Wayne's men erected in 1793 on ground

292

where American troops under General Arthur St. Clair had suffered a bloody, disastrous defeat by Indian warriors in 1791. Memorials maintained by the state mark both this historic site and the Fallen Timbers battleground.

A few miles downstream from the latter, on the opposite shore of the Maumee, is a third state memorial, on grounds once occupied by Fort Meigs. This stronghold, built by General W. H. Harrison, withstood two sieges by British and Indian forces in the War of 1812.

In addition to General Wayne, there were many other heroes in critical frontier days whose names are honored in the annals of the region. Two among these, Peter Navarre and Peter Manore, resourceful, courageous men of French descent, were prominent figures in northwestern Ohio phases of the War of 1812. Because both had many friends among the Indians, spoke their tongues, and understood Indian nature, they were able to aid the scattered settlements in the lower valley effectively and to preserve many a white man's scalp intact. Toledo has a Navarre Avenue and a Navarre Park, both named for Peter Navarre. The restored cabin in which he once lived is preserved on a site at the Toledo Zoo.

Whites, befriended signally by Manore, failed to show any substantial appreciation. But the Indians, remembering his kindnesses to them, ceded to him a large tract of land in what is now Providence Township, Lucas County. There, on the Maumee opposite the present town, Grand Rapids, he founded the village of Providence. Fire and cholera struck the settlement with devastating effects in the early 1850's. The ghost town Providence, Providence Park, the township, and its Manore Road now stand as memorials to this stouthearted pioneer.

Peter Manore heard from the Indians a legend that has made a high rocky outcrop on the bank of the Maumee in Waterville famous locally as the scene of an odd incident that antedated the white man's coming. As the story goes, a boy among a band of Ottawas encamped near the spot, known as Roche de Boeuf, fell from the cliff and was killed on rocks below. The father, holding the lad's mother responsible, pushed her off to her death. Her next of kin, following tribal custom, brought the same fate to the father. The kinsman in turn met a like death. And so it went, until chiefs intervened and ended the slaughter. The following day, bodies of the victims, numbering two-thirds of the group, were taken from the river and given ceremonial burial.

General Wayne's troops, en route from Fort Defiance to an expected confrontation with the Indians, hastily erected near Roche de Boeuf a stockade that they named Fort Deposit. According to a tradition that has long persisted, some of the soldiers, aware of the fate that might lie ahead for them (the battle of Fallen Timbers was to come only a short time later), buried money and other valuables at the spot. In ensuing years the field in which the fort stood has been dug over repeatedly by treasure hunters, generally without success.

Several miles above Roche de Boeuf lies Girty's Island, so named because Simon Girty once lived there. Simon was one of four brothers, all of whom, as children, had been adopted into Indian tribes. All became notorious renegades, aiding the Indians in many forays against whites, often proving themselves more brutal and barbarous than the savages. Simon, exceedingly heartless and cruel, was guilty of innumerable atrocities against white men, women, and children, in Ohio, Pennsylvania, and

Kentucky areas. Yes, as in the Westerns now often viewed on movie and television screens, there was a full quota of bad guys in those days, a great many of them white.

Vanquished and outnumbered, a good many of them reduced to vagabondage and chronic drunkenness through the chicanery and greed of unprincipled white traders, the Indians were long since forced to leave the region. Little is left to remind us of them except the names of a few such chiefs as Pontiac, Tecumseh, and Little Turtle, and a handful of geographic names, including Erie, Maumee (a somewhat corrupted form), Ottokee, Ottawa, Sandusky, Seneca, Shawnee, Tontogany, Wapakoneta, Wauseon, and Wyandot.

As Americans in so many other places have done, citizens of Allen County, in the Black Swamp, named a town LaFayette, in honor of the great French general. Other names of French origin and a number of German names, all having in some degree historic significance, are found on maps of the region. The name of Wilshire, the first Van Wert County town, is reminiscent of a singular bit of experience in the life of its founder, Captain James Riley. After shipwreck, attended by great personal distress, off the coast of Africa, Riley was enslaved by Arabs, at whose hands he suffered acutely until a man named Wilshire effected his release. As an expression of gratitude, he gave the name of his benefactor to the town he laid out in the wilderness in 1822. The three Revolutionary War soldiers who captured Major André, the British spy, are memorialized in the names of Paulding, Van Wert, and Williams counties.

The hamlet of Junction, in Paulding County, was so named because at that point the Miami and Erie Canal, extending northward from Cincinnati, joined the Wabash and Erie Canal, connecting points in Indiana to Lake Erie. The town of Delphos started with the name "Section Ten" because of its position on the Miami and Erie Canal. Seven or eight miles westward lies the little town of Middle Point, named thus because it is midway between Delphos and Van Wert on the line of what is now the Pennsylvania Railroad. For many years a limestone quarry, probably the largest of the many quarries in the region, was operated at Middle Point. The observer finds it difficult to believe that the now abandoned pit, enormous though it is, could have yielded the untold thousands of tons of stone taken out for railroad ballast, for construction, and for macadamizing hundreds of miles of public roads thereabouts.

Toward the end of the last century, Middle Point could boast a normal school where young men and women from the surrounding area were educated as teachers and where prospective doctors and lawyers obtained their pre-college training. In about the same period a similar institution was established at Ada, in Hardin County, and a third, known as Crawfis College, was founded in a rural environment a few miles north of Pandora. The school at Ada expanded from time to time, becoming finally Ohio Northern University. Crawfis College was abandoned as an educational institution years ago. The building that housed it, after a long period of vicissitudes, was finally destroyed by fire.

Traversing the highway built over that portion of the outer beach of ancient Lake Maumee that extends eastward from Fort Wayne, through Van Wert and Delphos and beyond northeastward, the observant traveler will note in adjacent plowed fields narrow zones of yellow sand, varying in width and sloping gently away from either side of the highway. (Similar expanses of sand occur along roads built over intermediate beaches.) As this is a part of the Lincoln Highway, Ohio-U.S. Route 30, an important east-west road, it is much used, and the pace of most vehicles is rapid. License plates issued by states other than Ohio are common, those from Indiana naturally predominating. Practically every town on the route has one or more towering grain elevators adjacent to rail lines; this is country that produces cereals and soybeans on a large scale. High water-storage tanks are also conspicuous, but tall stacks emitting clouds of black smoke are comparatively rare—electricity now turns many of the wheels of industry. Current may come from local or nearby generators, but all of them are likely to be tied in with the interlocking network of transmission lines that crisscross the country.

At Van Wert the road deviates slightly from the line of the old beach to pass through the heart of the town. At the time when state and federal highways were being projected, many municipal bodies exerted themselves to make sure that the routes would pass over their principal streets. In recent years some of the towns and cities that succeeded in these efforts, now quite disillusioned, have been working with might and main to have outside traffic diverted from their streets. This is being done rather often because, in spite of the high costs involved, highway officials are eager to get away from the congestion caused by narrow streets, with traffic signal systems that not infrequently are so operated that they impede traffic flow.

On the highway at the eastern edge of Van Wert is

297

the Marsh Foundation, provided by a bequest of a local businessman, for the education and vocational training of boys and girls. A mile eastward is a branch of Starr Commonwealth. Nearby are two or three exceptionally opulent-looking farms, apparently developed as showplace hobbies. A little farther eastward lies the county farm with its large, old-fashioned brick house and commodious barn. Nearby is the burial ground, last resting place of the homeless and the indigent of several generations. To most local people, this establishment for years was the "poorhouse." Now it is known by the more euphemistic name of "County Home" or "County Infirmary."

The Lincoln Highway, bordered right and left by prosperous well-kept farms, has a full quota of automobile service stations, wayside restaurants, drive-in movie theaters, motels, and much-used state parks, now called "roadside rests." Fronting upon it are several rural churches in regular use. One of the oldest of them has an old-fashioned churchyard. Unlike other cemeteries that may be seen from the road, it has been so much neglected that ragged grass and weeds surround the monuments, some of which are broken, some toppling.

As in many other parts of the state and the nation, the construction of interstate highways goes on apace in the region. These roads, with overpasses for grade separation at secondary roads and railroads and with somewhat complicated clover-leaf interchanges at main highways, are extremely costly. It is an impressive fact that they are also taking out of production an enormous aggregate acreage of excellent agricultural land.

Local roads in several counties are identified by numbered or lettered signs at intersections. This system, doubtless useful to public officials, does little for the guidance of the stranger. Other counties name roads for pioneer families; the traveler may be unfamiliar with the names or their local historical significance, but the signs are likely to be much more helpful to him. Roads extending from Van Wert are commonly known by the names of the towns to which they lead: Decatur, Mendon, Middle Point, Rockford, and so on. A few miles from our farm on the Mendon Road, the Middle Point Road passes over "Walser's Hill," a rise so slight that it has no perceptible effect on a car's motor. Each crossing provokes a chuckle as I recall bicycling days when, to all riders pumping up the incline, it seemed a steep, high hill.

A good many narrow bridges, built in pre-automobile days, are found on rural roads. One of the two known covered bridges in the region was built about 1845 to span a creek some six miles south of Fremont. Not long ago the road there was rebuilt to bypass the bridge, which, closed to traffic and refurbished, with a convenient parking space near it, has become a popular tourist attraction.

Complexities of the megalopolis are seen developing along highways radiating from some of the larger cities. On either side of such roads may be seen good modern dwellings, often in spacious grounds tastefully landscaped. In some cases, commercial or industrial establishments stand cheek by jowl with the houses. Here, obviously, development came before the areas were zoned. Zoning regulations are now in force in many urban areas; they have also been adopted in a number of rural sections.

In and about numerous cities and towns, extensive suburban sprawls are in evidence. Here, quite often, one sees groups of boxlike, one-story houses, covered by low-pitched roofs, with little to distinguish one from another. To make matters worse, these prefabricated houses in too many instances are crowded closely together on small lots. Unattractive as they usually are, few commercial developers would consider putting up any but such "ranch-type" houses because they sell very readily. It is to be expected that in time the popularity of this style will die out, just as did the overpowering craze for "bungalow" house styles of a generation or more ago.

On almost any moderately long trip in the valley, one or more deserted old houses, standing forlorn and a little ghostly, with appurtenant buildings in a state of acute disrepair, will attract attention. Most of the glass is likely to be missing from windows— hoodlums and thoughtless urchins seem to feel that some unwritten law licenses them to work such destruction once a building has been abandoned. Timbers of the framework may be exposed, reminding the observer of the bleached bones of some great, long-dead animal. To emphasize the desolate aspect, a clutter of wreckage and junk commonly lies about in a jungle of rank weeds, among wild, unkempt trees and shrubs.

Why are houses left thus to decay and ruin? Quite probably answers to this question would show a wide variation. Generally, some hard economic facts, expressed or implied, could be expected to appear among them. Houses dating back a century or more were built of hard native wood fastened in place by cut-iron nails that, though rather brittle, nearly always hold in the old wood with amazing tenacity; however, the heavy frame timbers may be no longer level or plumb. Remodeling and modernization, therefore, would involve so much time-consuming labor that, at prevailing wage rates, the cost might be prohibitive—merely wrecking such a building can be a difficult, costly job. Now and then a company of local firemen, with the owner's permission, will set fire to an old house considered to be beyond further use, then proceed to try out various fire-fighting techniques.

Bricks for some of the older houses and other buildings were formed and burned at the construction site. At the time, that was a logical procedure because both the clay and the wood for burning it were at hand for the taking; the use of bricks from the nearest brickyard—if there had been any—would have required heavy hauling over roads that were extremely poor. The walls were laid up solidly, their thickness not less than the width of two bricks.

Ditching machines are busy during much of the year, even in a few cases slashing through growing crops in fields. Yet they never catch up with drainage work; clay or concrete tiles, piled in long ranks, are seen awaiting the ditcher on many farms that appear to be already fairly well drained. Giant self-propelled combines and cornpickers play brief but important roles at harvest time. On every hand are tractors, large and small, handling a surprising variety of jobs. I was told of a retired farmer who, having been refused a license to drive an automobile because of his advanced age, blithely took to his tractor for short sightseeing trips and for visits to neighboring farms. Every farm family has at least one automobile. A few keep a saddle horse or two; the majority of riders seen are girls.

The time was when corn shocks dotted many fields in fall and winter and a strawstack stood close to nearly every barn. These have long since vanished from the scene, as have miles of fences. All told, there are a good many well-kept silos, not infrequently two or more on a single farm. Occasionally, one may also note a silage pit with its plastic cover fastened down securely. The stored silage is fed not only to dairy cows but to herds of feeder steers, shipped in from Western ranges to be fattened through the winter. Here and there are auction barns where, at stated times, livestock is sold to shippers, to local butchers, or to farmers.

A few farms have buildings designed for the commercial production of poultry and eggs, sometimes with facilities for large-scale dressing, freezing, and storing of carcasses of chickens, turkeys, or ducks. At moderately frequent intervals one finds on farms groups of metal bins in which are stored grains as security for government loans. In numerous barnyards are pole-mounted electric lights whose greenish glow illuminates a broad area throughout all or part of the night, facilitating work about the premises and discouraging prowlers. These lights, gleaming from a score or more of farmsteads, near and far, give to a night landscape a strange, fairyland aspect. Many farms enjoy the benefit of protection against fire under contracts between township officials and fire departments in nearby towns.

Although timber is scarce, a trip in any direction is likely to bring into view at least one sawmill with an accumulation of waiting logs, eight to twenty-four inches in diameter. Despite the fact that electric interurban railways, numerous in the region at one time, were long ago abandoned and dismantled, one may now and then come upon a short section of graded roadbed or an idle old bridge spanning a stream. Piles of boulders, collected from fields, are seen on a few farms; no doubt most others have similar collections where they are not visible from the highway.

Hundreds of Black Swamp farms have fields of good alfalfa from which three or four cuttings per year go to nearby mills to be kiln-dried, pulverized, and bagged for use in feed mixtures. The grass, as it is cut, is blown directly into deep trailer-mounted boxes in which it is hauled to the kilns. Many are convinced that the harvesting equipment used has a more telling effect in keeping the pheasant population low than even the numerous foxes that devour countless eggs and prey upon the birds at all stages of development. It is said that in early summer the bodies of a score or more of incubating females, mangled by the machine, may be blown into a single load of grass.

Several of the country's largest food-processing concerns have plants in the region. At the height of the season each year they operate day and night, converting into canned and bottled products the many thousands of tons of tomatoes brought in from surrounding farms. It appears that the failure of Cuba's communist-directed sugar industry has greatly stimulated the production of Ohio sugar; sugar beets are now grown extensively by Black Swamp farmers and trucked to refineries that hum throughout the fall and much of the winter. Toward Lake Erie, large acreages are planted to cabbage, for conversion into sauerkraut, and to cucumbers, for pickling. Establishments for processing and packing both are maintained at central points. Great quantities of cherries from numerous orchards are packed by commercial canneries in the area. Large-scale production of all these specialized crops requires the help of a small army of migrant workers.

Let us now take leave of the valley by considering briefly its present aspects as a product of glacial action. No one can doubt that the Maumee River, in the course of its long life, has gradually grown larger or that its channel, islands, rocky rapids, and terraces have been subject to many alterations. Toward its mouth, principally to the westward, occur deposits of sand accumulated in the final basin of dwindling old Lake Maumee and left behind when the last of its water trickled away. No important changes appear to have occurred here except that in some places winds have piled the sand into drifts—it is known locally as "blow sand."

Traveling southward or southeastward from the river, one crosses the old lake beaches and several glacial moraines, all clearly identifiable. It is evident that the beaches have been modified to some extent by road-builders and farmers. Clearly, the low ridges of the moraines have undergone a great deal of wear and tear; they are destined to become less and less conspicuous because farmers commonly work over them in fields in a manner calculated to level them. Weathering and erosion—active from the beginning—continue, of course, affecting beaches, moraines, and other features, with far-reaching results that can by no means be ignored.

Notwithstanding the observable alterations, all of them together relatively inconsequential in the vast ensemble, it may be said that the valley remains fundamentally unchanged, its immense drift sheet, its characteristic topographic features, and its inherent qualities essentially the same as they have been for thousands of years.

CPSIA information can be obtained
at www.ICGtesting.com
Printed in the USA
BVHW081425200620
581930BV00001B/24

9 780814 207345